Recombinant DNA Research and the Human Prospect

Earl D. Hanson, *Editor*

A Sesquicentennial
Symposium of
Wesleyan University,
Middletown, Connecticut,
March 4–6, 1982

AMERICAN CHEMICAL SOCIETY
WASHINGTON, D.C. 1983

Library of Congress Cataloging in Publication Data

Recombinant DNA research and the human prospect.

1. Recombinant DNA—Congresses. 2. Genetic
engineering—Congresses. 3. Human genetics—
Congresses.
I. Hanson, Earl D. II. Wesleyan University
(Middletown, Conn.) III. Title: Recombinant D.N.A.
research and the human prospect. [DNLM:
1. Recombination, Genetic—Congresses. 2. DNA,
Recombinant—Congresses. QH438.7 R3115 1982]
QH442.R414   1982   574.87'3282   82-20613

ISBN 0-8412-0750-X
ISBN 0-8412-0754-2 (pbk.)

PRINTED IN THE UNITED STATES OF AMERICA
Second printing 1984

Contents

*Science-based technology has eased and
enriched the personal lives of billions of
humans, albeit in varying degree. No end to
that process is in sight. It has been the
great technological triumphs ... that
engendered the large-scale public support of
science. Our lives are pain-free and rich in
experience beyond the imaginings of the
past. That technology has also quickened
the pace of history—for good and for ill. And
it happens before our eyes.*

Philip Handler (1917–1981)
From "The Future of American Science"
The first Admiral Charles A. Davis Lecture
Naval Postgraduate College, Monterey, California

November 6, 1979

Philip Handler had agreed to be the lead-off
speaker for this symposium. To our deep
regret, his untimely death on December 29,
1981, precluded that participation.

Preface

MORE THAN 100 YEARS AGO Charles Darwin changed forever our view of ourselves. No living things, human beings included, said Darwin, survive unchanging; change is inherent in the history of life. Now we face the ultimate in that awareness of change: we can change ourselves. Thus we have entered what is being called the Age of Intervention. The knowledge and technology subsumed under the name of recombinant DNA has made possible this ultimate intervention. Also referred to as genetic engineering, the new technology includes our demonstrated capability to produce purposefully new kinds of microorganisms, plants, and animals through manipulation of their genetic materials—by recombining their DNA. Human beings are included in the potential of this research, and in the summer of 1980 the first tentative experiments in that direction were initiated.

What is the prospect now upon us? Are we playing God, as Freeman Dyson bluntly asserts? And, if so, is catastrophe inevitable, as he also asserts? Or is genetic intervention being done responsibly? Are we in control of the dangers and are we cautiously but surely extracting the human benefits of this research in terms of enhanced knowledge and social gains?

The prospect has an impact on our sense of humanity, on our social institutions, and on our scientific activities. Hence, the prospect can be analyzed from many perspectives—humanitarian, societal, and scientific. In this volume we examine what those people, especially the scientists, close to recombinant DNA research, are acknowledging as its effects on us and on certain of our institutions—scientific, educational, commercial, and legislative.

The symposium upon which this volume is based was presented by the Science in Society Program at Wesleyan University as part of that university's sesquicentennial celebrations on March 4–6, 1982. The symposium started with an introductory seminar of four papers aimed at providing background ideas, issues, and information, including vocabulary, for those unfamiliar with recombinant DNA research but desirous of learning more. Members of the Biology Department at Wesleyan University presented those four introductory papers; their papers, as all in this volume, have been edited slightly from the original presentations in adapting them for publication.

The speakers in the main body of the symposium and authors of the remaining papers published here are the following: **Philip H. Abelson,** Editor of *Science,* Washington, D.C. Member of the National Academy of Sciences. A leading spokesman for the active role of science in human affairs and for effective national science policies. **George E. Brown, Jr.,** U.S. Representative from California. Member of Agriculture Committee and Chairman of the Subcommittee on Department Operations, Research and Foreign Agriculture with oversight on matters pertaining to recombinant DNA research. **Clifford Grobstein,** Professor of Biological Science and Public Policy, University of California, San Diego. Member of the National Academy of Sciences. Insightful critic of the interactions of science with society. Author of "The Double Image of the Double Helix." **Irving S. Johnson,** Vice President, Lilly Research Laboratory, Eli Lilly Company, Indianapolis. An active participant and industrial spokesman in public discussions of recombinant DNA research. **Salvador E. Luria,** Professor and Director, Center for Cancer Research, Massachusetts Institute of

Technology. Nobel Laureate and Member of the National Academy of Sciences. Leading researcher on viral and molecular genetics and their implications for medicine. **Robert A. Swanson,** President, Genentech, New York. A founder with Dr. Herbert Boyer of Genentech, a company pioneering in both fundamental research and marketable products from molecular biology.

It is a pleasure to acknowledge the special help of Max Tishler, University Professor, Emeritus, of Wesleyan University and former Vice President for Research, Merck and Company, in organizing the symposium and editing the papers in this volume. My thanks go also to Wesleyan University for funding this sesquicentennial symposium and for providing many essential support services. Finally, I wish to thank the Books Department of the American Chemical Society for their friendly and highly professional help in bringing these papers to a wider public.

EARL D. HANSON

Foreword

by Max Tishler

THE TWENTIETH CENTURY WILL LONG BE REMEMBERED as one with enormous change in humanity's capabilities, certainly the greatest and most rapid change in recorded history. Of the many discoveries that have brought about this transformation, splitting of the atom by Hahn and Strassman in 1938 looms over most others. What followed was a continual flow of new and associated developments: some useful, some harmful, and some that have given man the awesome ability to precipitate staggering devastation of life on this planet. A second discovery, probably of equal importance, was made by Watson and Crick in 1953 when they unraveled the structure of DNA, the molecule governing heredity. While a significant and measurable impact on society of what has followed has yet to be felt, clearly the effects of this discovery eventually will equal, at least, those associated with atomic fission. Yet a decade ago, we could not have foretold that recombinant DNA was a possibility and that humans may have new powers affecting life itself. But in 1973, when crucial experiments revealed to scientific audiences the possible powers of this new tool, almost god-like in scope, scientists close to the field of research became fearful of the potential harm if experimentation were left uncontrolled. Were scientists ready to handle the knowledge of recombinant DNA for the benefit of all of us? But unlike the scientists who participated in the Manhattan Project or who witnessed the first atomic explosion over the Alamogordo Flats, scientists of this generation, as a group, have a greater appreciation of the vanishing boundaries between science and human values. The now historic Asilomar Conference of 1975 on recombinant DNA held by a group of concerned scientists, many involved in recombinant DNA research, was a manifestation of this new responsibility of science to society. This conference provided a public forum for discussing the problem, and while it appeared to create an inordinate amount of needless and unsound fears in the minds of people, including politicians, it was a useful and productive experience. A tangible consequence was that the National Institutes of Health set up guidelines on the control of experimentation in the field of recombinant DNA. These guidelines are universally accepted by scientists in this country in both academia and industry.

The Wesleyan sesquicentennial symposium on the human prospect and recombinant DNA research was designed to assess both what has happened during the past ten years and the future prospects of DNA research. This seemed to be an opportune time to bring together people who are involved in and concerned with these developments for scholarly discussions in an academic atmosphere. A number of conferences on recombinant DNA have already been held, but nearly all have been narrow and restrictive in range. Certain conferences have dealt with the philosophical or ethical issues, others with control and regulation; still others have focused on commercial prospects of DNA research. For the Wesleyan symposium we chose to cover all major aspects of this development. It was our good fortune that we were able to convene a group of eminent people, each with different interests and experiences, who could cover the spectrum of its social implications.

It became clear from the presentations at the symposium and from audience discussions that the ability of scientists to handle and contain recombinant DNA research has been underestimated and that investigators are responsive to public concerns. While constant vigilance is essential, everything has moved in the direction of relaxing the tension and apprehension of a few years ago.

The field is continuing to expand and the excitement of the prospect of recombinant DNA in the health and welfare of the human race and other living things continues with no abatement. For the first time many substances essential for homeostasis can be made available in quantities and at costs unobtainable even through the miraculous synthetic capacities of today's organic chemists. Without recombinant DNA we might never have been able to produce enough interferon to establish its importance in living processes and to explore completely its utility in medicine, nor enough of a vaccine, which can be used on a grand scale with complete safety, to prevent hoof and mouth disease in cattle. And for the first time there is tangible hope for the control of genetic diseases and even some degenerative diseases. In my own mind, recombinant DNA has added completely new dimensions to the synthesis of complicated and essential natural substances. We can look forward with a degree of confidence to the creation of special microbes that will be able to produce a host of useful compounds, many extremely complex in structure such as antibodies, enzymes, genes, and viral peptides.

Another interesting topic of the symposium concerned the industry that recombinant DNA has spawned. This development has created a host of entrepreneurs and is reminiscent of the industries propagated by solid-state electronics. Large companies including many in the pharmaceutical industry and in the heavy chemical industry are also investing in these developments, and in many cases university scientists have become actively involved in these business ventures. It was stressed during the symposium that relatively few of these endeavors will survive and prosper. A note of foreboding raised during the discussion concerned the superb ability of Japanese microbiologists who today probably exceed their American counterparts in scientific publications.

The role of the university and its faculty in the applications of recombinant DNA and their melding with industry for exploitation of recombinant DNA products were examined by several participants. Recombinant DNA has brought the university into the maelstrom of the business and financial communities, and in some instances industry is creating a home on the university campus. While some universities in the past have been sensitive to the possibility of profiting from research carried out in their laboratories, recombinant DNA has opened the flood gates. The situation is easy to understand because the federal government, the major benefactor of university research for the past thirty-five years, is significantly reducing its financial support. Is recombinant DNA creating a precedent that threatens the tradition of the university as the cradle of research and the natural, nourishing shelter of freedom of inquiry? Will the commitments to mission-oriented research by the university and its faculty, motivated by profits, impair unfettered research and defile the tradition of intellectual freedom to pursue any path, wherever it may lead, into the unknown? The participants in the symposium were not particularly worried by this turn of events, but they left the audience with a feeling of need for constant surveillance and assessment—because society will lose its capacity to discover if the university does not remain free to explore its own ideas.

I came away from the symposium with the realization that a new age is here and with confidence that recombinant DNA will be pursued to benefit mankind. Trying situations will arise from time to time, but on balance humanity will triumph. Some of the hopes for this development may fall short of predictions; but let us not forget that the prophets of the past most often underestimated the progress of science. Many of the scientific developments that now characterize this era were never dreamed of fifty or even twenty-five years ago.

MAX TISHLER

1

The Seduction of Perfectability: Genetic Intervention Comes of Age

Earl D. Hanson*

Science begins in those unmeasured moments when human minds, often unself-consciously, set themselves a problem and proceed to unravel it by finding testable answers. And then science emerges among us as those tests are completed and when the necessary conclusions engage the fabric of human living. For genetics, as a science, the crucial emergence is conventionally and rightfully given to Mendel's work on inheritance in the garden pea, first published in 1865. It was curiously ignored until its rejuvenation in 1900. For the next three decades, there was a swelling tide of interest in the brilliant experimental work that reviewed, extended, modified, and, of necessity, reconceptualized Mendel's hereditary factors into genes that store, express, replicate, mutate, and recombine the biological heritage of all living things.

*Science in Society Program, Wesleyan University, Middletown, CT 06457

0—8412—0750—X/83/0001$06.00/0

Lessons of The sophistication of this aspect of genetics led smoothly and inevitably, in the next two decades, to biochemical genetics and that extraordinary fusion of chemistry and biology embodied in the deoxyribonucleic acid (DNA) molecule; more explicitly, the double helix of DNA now famous as the material basis of heredity. Since midcentury, molecular genetics has matured further, first in microbial systems and now also in multicellular systems where developments, inconceivable in the microbial genetics of viruses and bacteria, have further refined the Mendelian genetic paradigm or model to a position among the most elegant and challenging of all ideas in modern science.

When we now ask of that paradigm, "What have you really taught us?", we see that in having learned much about the nature of life we have also gained new visions of the power that inevitably accompanies knowledge. We see genetics transformed from Mendel's insightful experiments in the monastery garden in Brunn, Czechoslovakia, to multimillion dollar biotechnologies. We see how Darwin's sense of the changing nature of organisms is materialized in chance mutations inexorably selected by the necessary conditions of life. And this pinpoints the essence of evolutionary changes, changes that include human biological history, too. We see, also, how chemical programs in our cellular DNA specify much of what we are, and this knowledge gives rise to new challenges in medical treatment and profound challenges to our philosophies of what we ultimately conceive ourselves to be.

Age of Intervention The greatest challenge, for many of us, is encapsulated in Clifford Grobstein's phrase the *Age of Intervention*. This tells us that now, for the first time in history, we can purposefully intervene in the hereditary process. We can create, to suit

2

humanly defined purposes, new living things— microbial, plant, and animal—including ourselves. The prospect of human experimentation so as to intervene beyond the simple choice of whether or not to reproduce our kind has moved to the point of whether or not to produce a consciously designed certain kind of us. That arouses in any thoughtful person a critical stance regarding what is underway. This symposium is cognizant of that stance; it is part of the quest to understand and evaluate the implications that genetic intervention has for the human prospect.

We can create, to suit humanly defined purposes, new living things....

The urge to intervene genetically is compounded of two utterly human impulses. One is the urge to know and do more and this is largely the impulse energizing scientific and technological research. The other is the dream, as venerable as humanity itself, of improving the quality of our lives. This is shared by scientist and nonscientist alike but is perhaps more seductive for the latter.

As public and private individuals, most people strive to better their lives materially, biologically, intellectually and emotionally. As a society, humans gather their various resources to promote their perception of a common good. And as a world, we hope, however fitful and uneven our struggles seem to be, to make it a better place. Our great cultural traditions aspire to one form or another of perfectability. For the religious, it is some form of godliness and heaven on earth, predicated on a vision of some final form of Goodness. For those of a secular bent, there is a release from savagery, a striving for civilized order, and for some, even Utopian visions.

Our great cultural traditions aspire to one form or another of perfectability.

3

Seduction of Perfectability

The seduction of perfectability, of hope for some form of improvement or progress, elicits a basis for moral and practical support for those who seem capable of materializing that hope. In that context genetic engineering is the ultimate seduction. Although control over our environment is pervasive, it is also blighted in various ways. What remains as the ultimate challenge is to control ourselves, to intervene so as to minimize what we don't want and maximize what we do want. Human perfectability, attainable in some real degree within each of us, has been pursued, perennially, prior to the advent of genetic engineering. It has taken forms as disparate as moral exhortations and doctrinal pronouncements, as health foods and medical cures, as psychic serenity and mystic meditation. And now there is genetic intervention.

What remains as the ultimate challenge is to control ourselves...

Can genetic intervention be used wisely? There are those who dogmatically say no. If we presume to play God, they say, our ignorance and arrogance guarantee disaster. And there are those who enthusiastically urge us on to exploit cornucopian visions arising from genetic intervention. In between are the majority who say go slowly for wisdom is neither inevitably beyond human reach nor inevitably available just because we aspire to it. Wisdom is earned and the human prospects opened by recombinant DNA research are worth striving for because they do offer concrete opportunities to better ourselves, our society, and the world. But from history we know that such hard-earned prospects are never guaranteed. They only become attainable through the marvelous alchemy of human imagination, sustained effort, and a practicable vision of human dignity.

It is that prospect that is examined from various perspectives in the contributions that follow.

What Is DNA?

*J. James Donady**

Just what is DNA (deoxyribonucleic acid)? The answer to this question has become routine: DNA is *the* Genetic Material. The general public has been bombarded with this answer through the mass media. However, several questions about this genetic material need answers to appreciate more fully the potential power of recombinant DNA research.

If DNA is the genetic material, how can one type of molecule be the genetic material of all living species (plants, animals, and bacteria)? The answer is found in the *structure* of DNA. We will see that the DNA molecule is simple in its structural organization and yet versatile in the arrangement of components within that structure. Due to its versatility, DNA can carry the genetic information of *Homo sapiens* as well as virtually all other species. (The only exceptions—they are

*Biology Department, Wesleyan University, Middletown, CT 06457

0—8412—0750—X/83/0005$06.00/0

rare—are the species that use RNA as their genetic materials.)

How does DNA faithfully maintain the characteristics of a species in every individual of the species? The method by which more DNA is synthesized from the existing DNA is a copying process. The fundamental structure of DNA ensures that the copies will be identical to the original. Therefore, the *replication* of DNA ensures that every cell in each oak tree contains the complete genetic information of the oak and that oak trees will continue to produce oaks.

... the replication of DNA ensures that ... oak trees will continue to produce oaks.

How does the chemical information contained in the DNA produce the physical characteristics of the living organism? Such characteristics include the constant, species-specific traits, such as the long neck of the giraffe, as well as the subtle variations in traits, such as blue, brown, and other shades of eye color in humans. Here we will see that the chemical structure of DNA can be transferred to another, similar molecule (RNA). Subsequently, the RNA information is transformed into another class of molecules (proteins) that produce the characteristics of the individual and the species. These processes of information transfer are called *transcription* and *translation*.

In the following sections, DNA will be discussed in reference to its structure, replication, and transcription/translation.

The ABCs of DNA

The DNA molecule is linear and is composed of repeating units attached together like beads on a string. The repeating unit is called a *nucleotide*. Each nucleotide contains three parts: a phosphate group (P), a sugar molecule (S), and a nitrogen base (see Figure 1). There are four kinds of nucleotides contained in DNA. They are identical in their phosphate and sugar structure but differ in the structure of their nitrogen bases. The differences in the nitrogen bases ap-

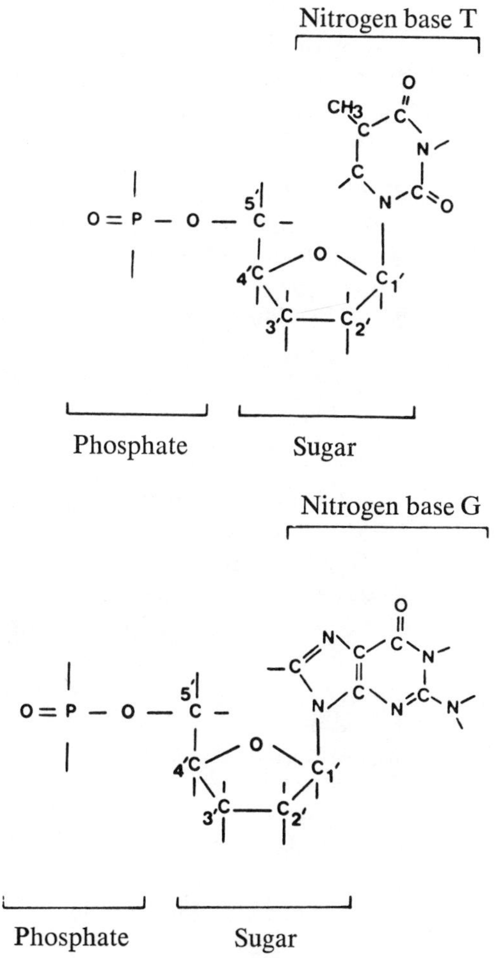

Figure 1. Nucleotides of DNA: Each nucleotide of DNA contains a phosphate group, a deoxyribose sugar, and a nitrogen base. There are four bases, thymine (top), guanine (bottom), cytosine (not shown), and adenine (not shown). Some atoms of hydrogen and oxygen are omitted. Carbon atoms on sugar are numbered 1' through 5'; C = carbon, P = phosphorus, N = nitrogen, O = oxygen, and H = hydrogen.

pear minor, such as a change of an oxygen atom for a nitrogen. However, these chemical differences do have consequence for the structure and function of DNA. The names given to these nitrogen bases are adenine (A), guanine (G), cytosine (C), and thymine (T).

Nucleotides are strung together by attaching the phosphate group of one nucleotide to the sugar molecule of another nucleotide. The phosphates attach to sugar at certain positions (3' and 5' in Figure 1) giving the linear structure a polarity; a 3' end and a 5' end. The nitrogen base of each nucleotide is not involved in the chemical bonding of nucleotides to each other along the length of the DNA strand.

The biologically active form of DNA is not one linear array of nucleotides, but two. The two linear strands of DNA are organized in antiparallel fashion. If one strand is oriented in a $3' \rightarrow 5'$ direction, the companion strand is oriented in a $5' \rightarrow 3'$ direction. The two strands are held together by interactions between the nitrogen bases (see Figure 2). Not all nitrogen bases can interact properly. In fact, only two nitrogen base pairs are structurally stable: adenine with thymine and guanine with cytosine. The nitrogen base pairs (A–T and G–C) form rungs of a ladder with the sugar–phosphate chains as the uprights of the ladder. Finally, the "ladder" of DNA is not straight. Each strand forms a spiral and the two strands form a double helix. For our purposes, we will ignore the helical structure of DNA.

The overall structure of DNA is simple; however, the possible arrangements of nitrogen bases on one strand are essentially infinite. The genetic information is contained in the linear arrangement of the four bases along one strand of DNA. Therefore, 3'AATTGCAT5' is different genetic information from 3'ATATGATC5' even though the same bases are utilized. This versatility in

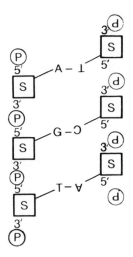

Figure 2. DNA ladder: The double stranded DNA molecule forms a ladder. The uprights of the ladder are formed by joining sugar (S) and phosphate (P) molecules at the 3'- and 5'-positions on the sugar. The sugar–phosphate strands are arranged in anti-parallel fashion. Nitrogen bases on sugar molecules pair to join the two strands. Nitrogen base A pairs with base T and base G pairs with C.

DNA allows the simple molecule to encode the genetic information of a human and a mouse. By utilizing differing amounts of the four bases and specifying the sequence of those bases, the genetic information of all living organisms can be stored in DNA.

Making Copies

The structure of one specific DNA determines the construction of faithful copies of its base sequences in a new specific DNA. In the double helix of DNA, each single strand is used as a copying template to replicate a new second strand. In an original ladder of DNA having a left side $(3' \rightarrow 5')$ and a right side $(5' \rightarrow 3')$, the left is used to make a new right side and the right side is used to make a new left (see Figure 3). The polarity of the template strand dictates that the new companion strand be of opposite polarity. The sequence of bases on one strand dictates the base sequence on the other: A dictates T, T dictates A, G dictates C, and C dictates G. This process, of course, reproduces the sequence of the original companion strand. In this way, faithful copies of the DNA base sequence and, therefore, the genetic information of the species are passed from cell to cell and generation to generation.

Transferring DNA Information

The genetic information stored in DNA can be transferred to other molecules and ultimately produces the characteristics of the species and the individual. The process of information

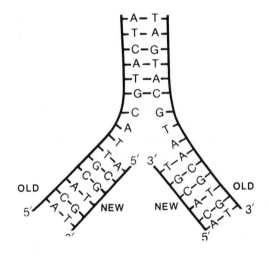

Figure 3. DNA replication: New strands of DNA are identical to old. The old strands are used as templates to dictate the antiparallel orientation of the new strands and A–T or G–C base pairs are required. Nitrogen bases of nucleotides are adenine (A), cytosine (C), guanine (G), and thymine (T).

transfer can be simplified to two steps, transcription and translation.

Transcription of DNA produces a similar molecule, called RNA (ribonucleic acid), which is more mobile and contains the genetic information from only a small fraction of the total DNA sequence. RNA is similar to DNA, but it differs from DNA in three ways: it is a single strand not a double helix, its sugar is ribose not deoxyribose (consequently RNA not DNA), and it utilizes the nitrogen base uracil (U) in place of thymine (T).

The transcription of RNA from DNA utilizes the same rules as DNA replication (see Figure 4). A portion of one strand of DNA is used as a template. The polarity of the DNA strand $(3' \rightarrow 5')$ dictates the opposite polarity for RNA $(5' \rightarrow 3')$. The base sequence in the DNA strand dictates the complementary base sequence in RNA (except A dictates U in RNA). The process of transcription transfers DNA information into RNA using the same basic language of nitrogen base sequence.

DNA	RNA
Double strand	Single strand
Sugar is deoxyribose	Sugar is ribose
Nitrogen bases are adenine, guanine, cytosine, and thymine	Nitrogen bases are adenine, guanine, cytosine, and uracil

The limited portion of DNA, which is transcribed into RNA, is a *gene*. Each gene contains a specific piece of genetic information. Since the information from a given gene is needed in only some cells or at only certain times, the transcription of limited portions of the total DNA is an efficient utilization of the total genetic information of the species.

What are the active molecules that produce the

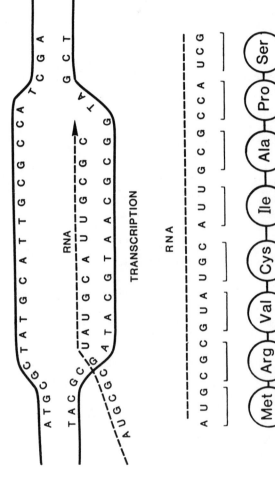

Figure 4. Transcription and translation of genetic information: RNA is transcribed from DNA using the rules of base pairing (A–U, T–A, G–C, and C–G). A three base genetic code allows the RNA to be translated into a string of amino acids to form a protein. Nitrogen bases: A = adenine, C = cytosine, T = thymine, U = uracil, and G = guanine. Amino acids: Met = methionine, Arg = arginine, Val = valine, Cys = cystine, Ile = isoleucine, Ala = alanine, Pro = proline, and Ser = serine.

specific characteristics of the individual and the species? Neither DNA itself nor RNA is active in producing a characteristic such as curly hair. The active molecules are *proteins.* Proteins are the major structural and functional molecules of all cells and organisms. Ultimately, the information stored and encoded in DNA and transferred to RNA must dictate the production of a protein. In this way, the information in a gene will be realized in the structure or function of the organism. This final dictation of a protein from RNA is called translation (see Figure 4).

As the term implies, translation is the transfer of information from one language to another, that is, from the DNA–RNA language to the protein language. Proteins are linear molecules like DNA and RNA. However, the units of proteins are amino acids. The sequence of amino acids in a protein is responsible for its activity. Therefore, the translation process transforms the specific nitrogen base sequence in an RNA into a specific amino acid sequence in a protein. In this manner, a gene can produce a characteristic of the species.

The translation of information from the DNA–RNA language to the protein language utilizes the *genetic code.* The machinery of the cell "reads" the bases in the RNA sequence ($5' \rightarrow 3'$) and builds a sequence of amino acids. However, there are only four nitrogen bases that can be used to code 20 amino acids. Therefore, a triplet code is the minimum number by which four bases can specify at least 20 amino acids; the code therefore has 64 unique three letter words. Some of the code words are used for punctuation (start and stop). Some amino acids are specified by more than one code word. In this way, a sequence of three bases on the RNA linear molecule (such as UCG or UGG) dictates the position of one amino acid (such as serine or tryptophan) in the chain of amino acids that will be the protein. This accomplishes the transfer of stored information in the DNA into active information in the protein (see Figure 4). Again, since the genetic code is adhered to strictly, the species' genetic information is faithfully translated into the species' amino acid sequence and therefore the species' characteristic is present in all individuals of the species.

... consider the subtle differences that exist ... in any species.... What is responsible for them?

Finally, consider the subtle differences that exist between members of any species. Those differences are our various individualities. Many of these differences are genetic. What is responsible for them? The rare change of a nitrogen base in DNA, called a mutation, will be faithfully replicated in subsequent copies of that DNA. Furthermore, in transcription and translation, the change in DNA can alter the amino acid sequence of the protein and thus change a certain characteristic. The result may be minor (a change in eye color) or major (a genetic disease). Sickle cell anemia is a potentially fatal disease which affects the hemoglobin in red blood cells of humans. The gene that causes the anemia has one nitrogen base changed as compared to the normal gene. This change produces a change of one amino acid in the hemoglobin protein. The altered protein cannot function as well as the normal protein and results in anemia.

Clearly the simple DNA molecule holds a great deal of power and potential.

Clearly, the simple DNA molecule holds a great deal of power and potential. The ability of science to analyze and manipulate DNA has removed the exclusive power and potential of DNA from nature and placed it in our hands. The formation of recombinant DNA molecules holds great promise and brings with it issues and responsibilities not faced by any other species in biological history.

Recombinant DNA Methods

*S. Steven Potter**

Recombinant DNA methodology is powerful. It allows the scientist to answer questions that were previously unapproachable and it allows the industrialist to market pharmaceuticals that were previously impossible to produce. On the one hand it promises a deeper understanding of the mechanisms of basic biology, and on the other hand it offers immediate products for the public good. And on the third hand, which we hope is imaginary, lurks the potential for trouble that accompanies many new technologies.

And on the third hand, which we hope is imaginary, lurks the potential for trouble that accompanies many new technologies.

Despite this great power, the basic concepts and methods of recombinant DNA work are simple. In general, the idea is to connect a segment of animal DNA (for example, human DNA) with a segment of bacterial DNA. This recombinant molecule can then be inserted into a bacterial

*Biology Department, Wesleyan University, Middletown, CT 06457

0—8412—0750—X/83/0015$06.00/0

cell, which will rapidly divide by binary fission to give billions of copies of itself, each carrying an exact replica of that original recombinant molecule. In some cases, these bacteria are then used as little factories to produce the protein encoded by the inserted DNA. And, in other cases, the bacteria are just used to amplify the animal DNA sequence. The many copies of the original sequence produced by the dividing bacteria are readily purified, thereby greatly facilitating studies of gene structure and function.

"Film Editing" Recombinant DNA

The cutting and splicing of DNA molecules are the fundamental methods of recombinant DNA work. These are accomplished by using special enzymes (protein molecules that regulate chemical reactions) that are isolated from bacteria. The restriction enzymes are the cutting tools, with each type of *restriction enzyme* cutting in a very specific way at a particular nucleotide sequence in the DNA molecule.

For example, the enzyme Hae III (from the bacteria *Hemophilus aegyptius*) cleaves DNA at the base sequence guanine, guanine, cytosine, cytosine (GGCC). That is, wherever this particular arrangement of bases is found, the enzyme will break the DNA molecule. There are literally hundreds of different restriction enzymes available now, with many different base specificities. By using different enzymes, singly and in various combinations, it is possible to precisely segment DNA molecules.

The DNA splicing tools are another class of enzymes called DNA *ligases*. These proteins connect the ends of DNA segments, by covalently sealing them together. By using these enzymes, we can take the DNA segments generated by restriction enzyme cleavage and join them back together in different combinations. For example, fragments of animal DNA can be linked to fragments of bacterial DNA.

A special kind of bacterial DNA, called a *plasmid*, is especially useful for molecular cloning. Plasmids are small circular DNA molecules with luxury genes (such as antibiotic resistance genes) that bacteria can live with or without in most environments. Plasmids are independent replicons, meaning they can replicate indefinitely inside of bacteria, and because of their structure (small double-stranded circular DNA) they can be easily purified from bacteria. This unique set of properties made them particularly convenient for molecular cloning.

Making a New DNA

The basic protocol involves first opening up, or making linear, the plasmid molecules by cleaving or cutting with a restriction enzyme. This plasmid DNA is then mixed with animal DNA that has also been cut, generally with the same restriction enzyme (an endonuclease). The DNA splicing enzyme, ligase, is added and animal DNA is linked to bacterial plasmid DNA. The resulting hybrid molecules, consisting of artificially recombined bacterial and animal DNAs, can then be inserted into bacteria by a simple procedure called transformation.

A single bacterial cell, carrying a single recombinant molecule, can quickly reproduce, yielding immense numbers of exact copies of itself (a clonal population). We can then ask these bacterial cells to produce an animal protein, such as insulin, or we can simply use the bacteria as little gene factories that produce large quantities of a particular animal gene, which can be rather easily purified back from the bacteria.

We can then ask these bacterial cells to produce an animal protein, such as insulin....

This ability to produce and purify large quantities of an individual gene is tremendously useful to the scientist trying to understand how genes work. Prior to the development of recombinant DNA methods, it was generally necessary to work with complex mixtures of DNA obtained from whole animal cells. Needless to say it was extremely difficult to understand how individual genes worked when forced to study literally thousands, or even hundreds of thousands of genes simultaneously. The problem resembled trying to learn the French language by listening to a recording of Paris made from a mile above the city. The roar of millions of Parisians talking at once simply would not be decipherable, and neither was the roar of thousands of genes working at once. We needed to be able to kidnap a single gene, to amplify its voice by making many copies of it, and then to study it alone in the test tube, isolated away from the rest of the gene population. Recombinant DNA methods now make this possible, and even routine.

The problem resembled trying to learn the French language by listening to a recording of Paris made from a mile above the city.

It should be emphasized that this is an elementary picture of recombinant DNA methodology It presents the essential aspects but leaves out many of the details that make the story more complete and hence lead to a better understanding of the current potential of the technology. For example, some workers now use yeast and yeast plasmids for molecular cloning, instead of bacteria. And many workers use bacterial viruses, which can also be recombined with animal DNA, in place of plasmids. And yet another molecular cloning system makes use of cosmids, which are artificially constructed hybrids that are part virus, part plasmid. Moreover, we have not considered the general problems of "shotgun" cloning, where all of the DNA segments of an organism are randomly cloned, and the chief technical difficulty is identifying the one clone in a million that is of interest.

The contributions ... are already impressive with much more yet to come.

The essential point is that these are widely available techniques for introducing the genetic material of one organism into another for purposes related to basic or applied science. The contributions to both areas are already impressive with much more yet to come.

Recombinant DNA: Controversies and Potentials

*Barry I. Kiefer**

If one adds up all the claims that have been made
for the potential of recombinant DNA technology
(RDT) over the past 6 or 7 years, it would appear
that we have in hand the magic necessary to
solve all of humankind's problems and provide
for the individual needs of the entire human
population. Well—not quite. Or at least not yet!
However, even by the most conservative estimates, the potential is absolutely mind-bog-
gling. Recombinant DNA technology is clearly
one of the most powerful tools invented by humans in their never ending quest for total control of the forces of nature for their own benefit.
Both the forging of such a tool and its subsequent use necessitate controversy; and controversy there has been, is, and will be.

The purpose of this paper is to introduce some
of the controversial issues that have surrounded
this research and its application, as well as

*Biology Department, Wesleyan University, Middle-
town, CT 06457

0—8412—0750—X/83/0019$06.00/0

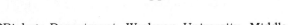

some of the realizable promises of the technology. It also provides an abbreviated historical and scientific framework in which these issues can be considered.

Three Cornerstones of Modern Biology

Biology's transition from natural philosophy to natural science took place a little over 100 years ago as a result of three revelations that occurred in the latter half of the last century. In 1859, Darwin published his study of the origin of species which provided extensive documentation that all species come from pre-existing species by a process of natural selection. In 1865, Mendel published his paper which outlined the fundamental laws of heredity. And throughout that period what has come to be known as the Cell Doctrine was being crystallized. It stated that all organisms are composed of essentially like parts, namely, cells and they come only from pre-existing cells. Life on this planet came to be viewed as an uninterrupted succession of cells with each aspect of this succession governed by processes or laws that are knowable.

To say that Darwin's work was controversial is to underplay its impact on the 19th century view of human beings and their universe. As any reader of the lay press knows, it is still controversial, but that speaks more to the roots of controversy than to the origins of species. To say that Mendel's work was controversial would be an extreme overstatement. It was totally ignored, only to be rediscovered in 1900 when the scientific community was better prepared to accept it. The science of genetics, then, is only 82 years old, which means that most of the geneticists who have ever lived are alive today. The relatively recent revelation of the universality of the genetic code is a prediction of these three ideas—individually and collectively. That tells us that all life forms on this planet speak the same language and, as in fields such as linguistics and an-

The science of genetics ... is only 82 years old, which means that most of the geneticists who have ever lived are alive today.

20

thropology, that is strong evidence for common origin.

The molecular genetics of today is the direct natural descendant of these three cornerstones of biology. We are about to reap a harvest sown at the turn of the century. The bounty of the harvest is in sight and what benefits it will bestow on us are still unknown, and, therefore, both exciting and frightening. The objective is the control of the genetic architecture of many of the individual life forms which inhabit this planet— including humans—for the benefit of our species. This is not necessarily the goal of individual scientists; it is the goal of our species, in the truly biological sense. The fundamental controversy, then, is which interest group will be at the controls, not whether the objective is correct.

We are about to reap a harvest sown at the turn of the century.

The controversies that have occurred in the last 8 or 9 years have included a variety of participants. Initially, there were two: scientists, pro and con. Then there was everybody, and now there are a different two: government and industry with the universities caught in the middle. And the emphasis has changed from "Is it safe and right?" to "Who's going to pay for and own it?" That shift was formalized when the Supreme Court ruled in June 1980 that new forms of life could be patented, thereby legalizing an unprecedented economic phenomenon: a multimillion dollar international industry had grown from nothing without a single product on the market. The entire, constantly expanding industry is built on a promise. No other fact so clearly summarizes the power of this technology. But before examining the potentials, let us review the origins and evolution of the controversies.

Roots of Controversies

In February 1975, an event unprecedented in the history of science took place. One hundred

21

and thirty-nine outstanding DNA researchers from 17 countries gathered together at Asilomar, Calif., to discuss their concerns for the possible unfortunate consequences of indiscriminate application of the newly developed recombinant DNA technology, to review the scientific progress in this area, and to propose appropriate ways to deal with potential hazards. This conference was the result of growing uneasiness first voiced at a Gordon Research Conference on nucleic acids in 1973 and sent in letter form to the National Academy of Sciences and published in the journal, *Science.* Philip Handler, president of the Academy, responded by asking Paul Berg, a pioneer in recombinant DNA technology (and Nobel Laureate in 1980), to formalize the biohazard concerns. Berg formed a committee of 10 and the fruits of their deliberations were put in the form of a letter published simultaneously in the three most widely read science journals in the world, *The Proceedings of the National Academy of Sciences, Science,* and *Nature* in July 1974, and publicized at a news conference called by the Academy. The concern centered around the possibility of inadvertently creating novel types of infectious DNA elements whose properties could not be predicted and which could be biologically hazardous.

The letter called for a voluntary moratorium on some kinds of recombinant DNA work until the risks could be assessed, even though there was no demonstrated risk at that time, and for the convening of what was to become the Asilomar meeting 7 months later. It also brought the recombinant DNA issue to the attention of the lay press who responded in the way to which we have become accustomed. The aftermath of Asilomar has been astonishing—the scientific community is still bobbing in its wake. While the conference resulted in the establishment of the National Institutes of Health (NIH) guidelines to contain the risk and ensure the safe unhindered continuation of recombinant DNA work, the social and political storm that followed was unanticipated and has obscured what was begun as a unique, socially responsible process, and has resulted, ironically, in an increased public mistrust of science.

In the years that followed, considerable experimental evidence together with the development of engineered strains of microbes that can only live under controlled conditions in the laboratory have made it clear that many of the biohazards originally feared simply don't exist, and the NIH guidelines have been restructured accordingly. But the fundamental question—What should be regulated by whom?—remains and the rapid commercialization of the technology raises again some of the biohazard concerns. There are no laws requiring private industry to comply with the NIH guidelines. In this context, it is worth remembering that the after-the-fact discovery of the unquestionable link between cigarette smoking and several clinical conditions including cancer has had little apparent impact on the economic growth of the tobacco industry. In addition, there are new concerns and new controversies. The questions pertain not so much to gene splicing as to profit splicing among universities, researchers, and private industry. And there are very real fears that the traditional defenses of scientific integrity are in danger. The changes in the relationships between academic and industrial sectors have been amazingly rapid and somewhat haphazard. American universities have been built on the principle of academic freedom. The research that goes on in them has been characterized by openness, free exchange, adherence to the tenets of the various disciplines, and a search for truth. Industrial research of necessity has been characterized by secrecy and a search for profit. How these different goals and ground rules can be merged successfully remains to be seen. Several papers in this volume will address aspects of these issues.

And there are very real fears that the traditional defenses of scientific integrity are in danger.

23

As in any case of a planned disturbance of the natural order, the negative possibilities must be evaluated in the context of the positive potential, which in the case of recombinant DNA technology appears enormous.

Genetic Engineering and Basic Research

Recombinant DNA technology is itself a fruit of basic research, most of which was supported by tax dollars, and genetic engineering *per se* is nothing new; recombinant DNA technology allows it to be done more precisely, more selectively, more predictably, and more extensively. The engineering can be subdivided further into direct manipulation of human genomes and everything else, because everything else has to do with having some other organism make or do something we need or want.

... genetic engineering per se is nothing new ...

The oldest form of genetic engineering is selective breeding. It is a major part of the agriculture industry and has resulted in among other things the absurd number of different kinds of dogs and cats we have. It is a "biotechnology" that is inseparable from the history of humankind. There have been astounding successes and forgotten failures; and there have been abuses, political, economic, and scientific. The history of selective breeding serves as a known background against which one may wish to consider the promise and perils of recombinant DNA technology.

The method is simple: one selects individuals with desirable traits and breeds them. Among the offspring, those that have the desirable traits of both parents are selected and crossed. This process is repeated until, with luck, a true breeding line with all of the desired characteristics has been established.

It was inevitable that a program that has been so successful in producing superior plants and animals would be suggested as a means to better the human species—the so-called science of eugenics. The rise of the science of genetics during the beginning of this century was paralleled by the eugenics movement. While the original scientific interest in the concept of eugenics may have been founded in honest, honorable, utopian desires to improve the human condition by the elimination of genetic deficiencies, the ultimate political impact of the eugenics movement was to provide support for racism and social oppression. Sterilization laws in 31 states, laws prohibiting racial intermarriage, and the Immigration Restriction Act of 1924 are some of the more obvious examples of sweeping social and political conclusions derived from unfounded theories of race improvement. The "scientific authority" for all of this legislation was provided by testimony from a few of the leading eugenicists of the period.

Clearly, one major reason for the success of the eugenics movement was the failure of the majority of geneticists to come forth and denounce the unfounded conclusions. We all know the extremes to which these ideas were put in Nazi Germany; less well known is that much of the racial and ethnic hate that exists in this country today has its roots in the eugenics movement and sterilization operations are still being performed in the absence of informed consent.

... an informed public is an essential ingredient in the decision-making processes that will mold the future of the human species.

This dark and too often forgotten chapter in the history of genetics serves to remind us that scientists have a social responsibility and that an informed public is an essential ingredient in the decision-making processes that will mold the future of the human species.

In principle, selective breeding among humans is not an absurd idea. There are two different possible approaches: so-called *positive eugenics* which tends to promote breeding among *phenotypically* desirable adults, and *negative* eugenics which attempts to restrict breeding among *genotypically* undesirable adults so as to diminish the burden of hereditary diseases. As for positive eugenics, we do not appear to be quite ready for a "computer-match" approach to improving our species. However, there are alternatives which, curiously, may be more morally and socially acceptable. Two are already practiced in certain sterility cases. Artificial insemination with chosen sperm is now a fairly common choice for couples in which the male partner is sterile. Human sperm banks have been in existence for several years. *In vitro* fertilization followed by surgical implantation of the resultant embryo into a host uterus has been a standard practice in the cattle industry for 20 years. The basic procedure is to obtain mature eggs from the desired female by either surgical or hormonal techniques, fertilize them in the laboratory with sperm obtained from the desired male, grow the embryo to the implantation stage, and then surgically transfer it to the uterus of a surrogate mother who brings it to term. There are now at least a few normal, healthy humans who have been brought into the world in this manner.

... the prospects for the cloning of humans have long tantalized imaginations and egos.

Since the aim of positive eugenics is the perpetuation of desirable characteristics, the prospects for the cloning of humans have long tantalized imaginations and egos. A procedure for cloning vertebrates was first perfected in frogs in a brilliant attempt to address a fundamental biological question concerning the genetic equivalency of all the different cells of an individual. The technique involves the withdrawal of a nucleus from a body cell followed by the insertion of that nucleus into a mature egg whose nucleus has been inactivated or surgically removed. The egg is then chemically or physically "activated," and the resulting embryo develops into a genetic replica of the individual who denoted the nucleus. While there is no evidence that this technique has ever been employed in humans, it has succeeded in mice. Unlike the original frog experiments, the main reason for doing it in mice was to show that it could be done. And if it can be done in mice, it can be done in humans. A concerted effort could yield a fairly routine human procedure within 5 years.

Had any of these techniques been available at

the time of Adolf Hitler's rise to power, it seems certain that they would have been used—*en masse.*

Negative eugenics aims at reducing the spread of deleterious genes throughout the human population. It is practiced today on a voluntary basis following genetic counseling and only with regard to the relatively few but growing number of conditions in which both the genetics and the disease are well understood and where carriers can be clearly identified.

Manipulating Human Genes

This brief and selective history of eugenics as a form of genetic engineering serves to introduce a consideration of what is possible or plausible for the manipulation of the human genome with recombinant DNA technologies. It is now possible to clone specific human genes whose dysfunction leads to serious clinical consequences. This has been done with several genes including the insulin gene as well as the genes for other human hormones. There are still major technical problems to overcome before the successful insertion of a human gene into a human host can be achieved. We must find appropriate vectors and ways of delivering the gene to the target tissue. The vector must be able to replicate, transcribe, and translate and must remain under the control of the host. Each of these represents major research problems, and we must be certain of each aspect, for once done, the process could be irreversible. Successful gene transfers in mice have been reported recently.

The current alternative to gene insertion is to provide the individual with the substance that his or her genetic machinery can't make, and that brings us to the "everything else" category of contemporary genetic engineering. This is the promise on which the industry that could conceivably be the dominant industry of the next century is currently being built.

In all cases what is being done is that a specific piece of DNA that codes for a specific protein is inserted into a microorganism that can be grown easily and in great quantities to produce large amounts of that protein product. That product can be isolated and purified or used directly by the microbe to achieve a desired result. Since virtually all biological reactions are mediated by specific proteins and biological reactions provide us with almost everything—the oxygen we breathe, the food we eat, and much of the energy we currently consume—the limiting factors in the application of recombinant DNA techniques for the good of the species appear to be human ingenuity and wisdom.

The Immediate Future

There are three areas that hold the most immediate promise: pharmaceuticals, chemicals, and food processing. In the pharmaceutical industry, recombinant DNA technology is bringing new approaches to the large scale manufacturing of antibiotics, hormones, vaccines, enzymes, and antibodies. Substances available only in minute amounts from natural sources now can be manufactured in large quantities. In addition to providing increased availability of and access to well-characterized medically important substances, this technology will permit in-depth exploration of substances whose potential is not well known, as well as allow for extensive clinical testing. An example of a pharmaceutical product whose development is well underway is human growth hormone. It is too large to be synthesized chemically and the current source, extraction from the pituitaries of human cadavers, cannot be expanded. Furthermore, unlike insulin for example, nonhuman versions of growth hormone have no activity in humans.

For the chemical industry, the impact of this technology will cut across a large spectrum of chemical groups: plastics and resins, flavor and perfume materials, synthetic rubber, medicinal

29

chemicals, and the primary products from petroleum which serve as the raw material for the synthetic organic chemicals. This technology also offers several advantages over the chemical processes currently being used to transform biomass into organic chemicals such as ethanol. Biomass is renewable and with proper management could provide a very reliable source of energy. The processes used are safer and themselves consume less energy. The microorganisms offer one-step production methods with less pollution.

The food processing industry also can expect significant changes in at least two areas: engineered microbes that transfer inedible biomass into human or animal foods, and microbes that aid in food processing either by acting directly on the food itself or by providing materials that can be added to food.

This abbreviated list must be considered only as an introduction to the future potential of the current technology. Several of these areas are addressed in detail in subsequent papers in this volume.

To date, the most important contributions of recombinant DNA technology have been in the area that spawned it, namely, basic research. In the preceding paper, Dr. Potter has outlined the tremendous impact of this technology on our understanding of the organization of genes and how they are regulated. It has also led us to unanticipated (and, therefore, extremely exciting) discoveries. For example, we now know that many genes contain small pieces of DNA inserted into the coding region of the gene. Some of these "insertion sequences" are mobile—the so-called "jumping genes"—and can insert in several regions of the genome and affect the functioning of specific genes. These discoveries have broad implications not only on theories of gene structure and function, but also on theories of both developmental and evolutionary mechanisms.

Finally, there are untold numbers of basic research problems in biology and chemistry that cannot be addressed simply because of lack of material. Recombinant DNA technology offers the opportunity to provide that material and to initiate an attack at the molecular level on long-standing fundamental questions relating to how living systems work. What the answers will be and where they will take us is unpredictable. And that has always been the real excitement of science!

5

Advent of a New Age of Intervention

*Clifford Grobstein**

Forecasting is a slippery business and I claim no special talent for it. Nonetheless, having spent my active intellectual life in what may prove to be one of the biologically most productive half-centuries in history, I am convinced that we may soon witness the advent of a new age of intervention, one that may be fully manifest early in the new millennium, now less than two decades away. In what follows I shall try to justify my conviction and to trace some of the implications it carries.

Human intervention is, of course, no new phenomenon. We are a meddlesome species. We spend virtually our entire lives meddling, intervening, and quarreling over the actual or possible consequences. We intervene in the lives of

*Program in Science, Technology, and Public Affairs, University of California—San Diego, La Jolla, CA 92095

0—8412—0750—X/83/0033$06.00/0

other people, of other nations, and of other species. The surface of the earth is scarred by our interventions, and we are beginning to leave traces on the moon and other planets. A half-century ago, intervention was a word used primarily in international affairs. Interventionists differed with isolationists in believing that rising conflict in far-away Europe and the Pacific would impact on our own way of life and that we had to intervene to exert influence on the outcome. The argument was terminated when catastrophe (and reality) struck in a "far-away" place, Pearl Harbor.

My subject is intervention in an entirely different sphere but its possibility and its consequences are no less quarrel inducing. We are now talking about intervening in, becoming involved with, and exerting influence upon the central process of all life: its hereditary component and the means of its expression in development. In the past half-century, and particularly in the past two decades, we have achieved a markedly enhanced capability for this kind of intervention. As a result, there is a perceptible thrust toward ever greater human control over reproduction, not only of bacteria, wheat, and cattle, but even of our own species.

Age of Reproductive Intervention

To speak of a prospective age of reproductive intervention does not mean that we have not meddled in the matter before. Reptiles existed before the Age of Reptiles and stones were probably thrown and even crafted before the Stone Age. It is the fact that genetic, developmental, and reproductive interventions are not only becoming far more common but far more precise and pervasive in consequence than ever before. They are becoming a dominant mode, applied with increasing subtlety and power, in the management and exploitation of economically useful species, whether microbial, plant, or animal.

Moreover, the biological commonality of mammalian reproduction is becoming ever clearer, raising the momentous possibility that human beings may apply interventions that succeed in mice, cattle, or monkeys to our own species.

A number of milestones lie along the course of this rising intervention, the onset has been gradual, an advent or coming rather than an event *de novo.* It is not insignificant that the term advent carries a religious connotation, as in the Second Advent of Christ. This fits the fact that the interventive options that now lie before us are deeply emotive as well as profound in social implication. They touch the core of personal individuality, of ancient traditions, and of strongly held faiths. It is not surprising, therefore, that they evoke doubt and trepidation in many quarters, especially strongly among those who adhere firmly to fundamentalist religious traditions.

However, the capability to control and manipulate hereditary and developmental processes can also be conceived to hold forth great opportunities and benefits. Given such contrasting expectations of consequences, the new age is not likely to materialize without controversy and turmoil. My rationale in highlighting these emerging options is the belief that frank acknowledgment of the advent and the tensions that are generated may ease the process of accommodation and sound decision—as slow creep at the surface of the earth is believed to release energy that otherwise might be discharged in a disruptive earthquake. We badly need discussion that anticipates events and necessary decisions. There is no better arena to initiate deliberation of this kind than in universities dedicated to rational discourse.

Having set the stage with generalities, what is the substance of the matter? A confluence of research currents has created the platform for this new age of intervention; currents from genetics, epigenetics, cell biology, molecular biology, and endocrinology in particular. A major current, of course, is the remarkable dynamic thrust of molecular genetics. This led to the identification of DNA as the material basis of hereditary information, the demonstration of its relatively simple but trenchant structure for replication and transmission between generations, and its capability to add to essential chemical stability just the proper amount of the spice of error and resulting variability. Beyond this is the extraordinary manipulability of DNA that has created the illusion that all of the complexities of life are constructed out of a marvelous tinker toy.

The power of these generative insights into the molecular basis of heredity is far from spent. Every month seems to bring new significant advance. The manipulability of DNA is enormously stepping up the pace of innovative interventions in genetic systems. Between recombination of existing DNAs and the synthesis of both old and new ones, the range and depth of control of genetic processes can be conceived to be practically boundless. Bacteria, among the most ancient of earth's lineages, have been altered so as to synthesize substances that took eons of evolution beyond their own origins to invent. Thus, the capacity to synthesize the component products of organisms of increased size and complexity has been grafted into ancient simple organisms with the primordial capability for rapid reproduction. This wide heterochronic hybridization is establishing the base for new biotechnological industries of expected enormous importance in such fields as medicine, energy production, agriculture, environmental management, and even financial management. An early facetious suggestion of mine that the first applied use of DNA might be to paint it on dollar bills was not far off the mark. A necessary caveat is that, for the moment, most of this is more imagined than accomplished. How much of it will actually occur must still face the stern test of reality and hopefully also careful human decision.

Certainly this is true of the particular aspect on which I will concentrate: the possible application of this new knowledge and set of technologies to the human species. The basic chemistry of DNA is the same in humans as in virtually all other forms of life. This means that DNAs not only from humans but from other life forms are theoretically combinable and that genes either from other species or newly synthesized, might be inserted into the human genome. The possibility is strengthened by recent reports of directed genetic change in cultured mammalian cells and in early mammalian embryos. In the latter case, there has been a limited number of instances of the directed change being transmitted to a succeeding generation, indicating passage into and replication within germ cells.

The basic chemistry of DNA is the same in humans as in virtually all other forms of life.

Once again, sound perspective requires recall that genetic modification even across generations is not without precedent. Assiduous selective breeding, enhancement of mutation rates, cell fusion, and other such tricks have long been powerful tools to produce new agricultural and laboratory stocks of many plant and animal species. Less dramatically, social traditions, migratory patterns, various forms of contraception, and technological means to control disease and other threats to survival undoubtedly have influenced the genetic base of human characteristics. However, the new capability for genetic intervention can yield far more precise and deliberately designed change, whether by transfer of genes already established in other species or of genes specifically constructed for a particular purpose. Since these procedures are at least theoretically applicable to all species, possible evolution by human design has been increased significantly in range and depth.

Nor is there any *biological* reason why the human species should be regarded as immune to such intervention. One effort in this direction has already been recorded, that is, to alleviate the hopelessly tragic state of advanced β-thalassemia induced by genetic defect by treating with DNA for normal hemoglobin. The effort was futile and widely denounced both as premature and as a violation of current regulations. Nonetheless, it illustrates the strong motivation among clinicians to find more direct and successful solutions for more than 2000 diseases that are known to result from various genetic defects. Most of these defects shorten life, decrease the quality of life, and involve great financial and emotional cost to families and societies. The medical rationale for genetic therapy therefore is high. Moreover, the correction of genetic defect in somatic cells by gene insertion does not, in itself, seem very different from efforts to cure cancer through chemotherapeutic drugs or radiation, well known to produce powerful and far less specific genetic effects. Gene insertion, if feasible, should be more direct and certain as a corrective agent.

β-Thalassemias and certain other blood disorders have a relatively simple genetic base in one or a few genes occurring as a complex, that function at least primarily in one cell type, the red cell. Beyond these are genetic pathologies of more complex origin, e.g., Down's Syndrome where the defect lies at the level of a whole chromosome. Whether these can or should be dealt with through genetic therapy raises still more difficult technical and policy questions. The ultimate trajectory of the examples is reached, however, with the possibility of genetic intervention that is aimed specifically at modifying germ cells and hence future human generations.

The long conjectured capability to accomplish genetic intervention was brought a significant step closer technically with the achievement of external *(in vitro)* human fertilization in 1978. This new reproductive intervention has as its medical rationale the relief of infertility in women due to blockage of the oviducts or Fallopian tubes. Successful use of the procedure has now been reported in England, Australia, the United States, and possibly India. The number of clinics offering the procedure is increasing rapidly. Available literature suggests that some 14 clinics may be operating or coming into operation in a half-dozen countries, and something approaching 20 babies already have been born following external fertilization. The procedure must still be regarded as in clinical trial but the level of demand and slowly rising success rate suggest that it will achieve the status of an established medical practice before too long.

What is important about it in relation to the new age of intervention is that it opens a window on a previously almost totally inaccessible period of early human development. Internal fertilization normally occurs in the upper reaches of the oviduct and the early embryo takes several days moving down the oviduct to the uterus. Development is presumed to continue in the uterine cavity for some unknown period before implantation in the uterine wall. Prior to successful external fertilization, embryos were recovered for observation during pre-implantation stages only under highly unusual and unplanned circumstances, and the number was very few. However, for clinical success of the new procedure, the externalized egg is fertilized in a laboratory dish and observed at intervals for 2 to 3 days prior to insertion into the uterus. This assures that the egg is not grossly abnormal and allows the uterus to reach physiological readiness for implantation. It also provides the "open window," whether for observation or intervention.

Intervention in Early Human Development

The developmental human stages thus observed and opened to technical manipulation are precisely the ones that have been shown in mice to accept and express inserted DNA. Moreover, as already noted, there are early reports that the DNA finds its way into germ cells since it is detectable in a succeeding generation. The accessible human stages are also those in which, in mice, such manipulations have been successfully carried out as nuclear removal and insertion, direct replication of haploid chromosomes (diploidization), and embryo fusion. It is also the stage that in other mammalian species has been cultured and continued in development to early organogenesis (partial ectogenesis). Speaking biologically and technically, the open window thus offers wide options for substantial interventions in human heredity and development.

Questions in Human Heredity and Development

Having said this, one should immediately call attention to the normative questions that are raised. It is often said that what technologically *can* be done *will* be done. Stated that baldly the statement is clearly refutable. For example, it is now technologically possible with nuclear weapons to destroy much, if not all, life on earth. We have not done so yet. It was technologically possible to produce a commercial SST; the United States did not do so and the European effort may soon be terminated. Use of various pesticides has been curtailed or foregone. Rather than cite further examples, it is perhaps better to restate the proposition more usefully: what can be done technologically will *sooner* or *later* be done if there is *net advantage* to be gained by those who *make the decision*. In this form, the proposition is less dogmatic and more applicable to the options that we are discussing. In particular, attention is focused on the questions when, for what and whose advantage, and by what decision process?

The logical progression of the technological scenario, assuming no other obstacle, points toward substantial interventions in human heredity and development, conceivably yielding some measure of control over human evolution. Such a prospect for various reasons is disturbing to many people.

How likely is this technological scenario and how soon will decisions be needed?

What advantages are likely to be sought at each step in the scenario and who will gain them?

What disadvantages may be foreseen and who will suffer them?

Who will make the decisions?

What impetus or constraints will be imposed by human values and formulated purpose?

Will the process be shaped by such human factors or is it governed by some other inexorable determination, e.g., an indomitable inner engine of technological thrust or some even less controllable higher historical necessity?

These questions raise explicitly and sharply the role of conscious human purpose, whether of individuals or societies, as a determinant of technological advance. More specifically, they confront what is often called the moral issue: whether human purpose should have any standing in determining the human evolutionary future. They also highlight the issue of current capability to make sound collective choices among objectives that involve the interests not only of those now alive but of future generations. These issues arise and have been discussed in connection with other technologies, for example, strategies of warfare, of energy development, and of environmental management. The issues are particularly poignant when they relate to possible self-determination of changes in humanity itself.

41

Is the technological likelihood of humanity influencing its own evolution high enough to warrant concern and serious analysis at this time? Estimating the likelihood and its degree is itself a difficult technical problem. However, even without being able to specify just how high the likelihood is, two considerations suggest that the question is timely. The first is that, for reasons already stated, the likelihood is certainly higher than it was 50 years ago, at a time when the apocalyptic philosophy of national socialism projected racial purification through controlled breeding. Today, a comparable world outlook could find in recent scientific advances a seemingly more promising technical path to its objective than the primitive and puny means available a half-century ago. A second reason the question is timely is that such conjectures of malign use can create enough anxiety to constrain further advances in knowledge that could be beneficial to health or improve available resources. Since the technological likelihood is already high enough to generate imagined outcomes with specific social consequences, it is time to assess it more fully and to sort out what is realistic from what is speculative. Such sorting is essential to relieve immediate decision making of irrational fears inspired by worst case, long-term speculative scenarios.

Realistic Questions for the Near Future

Three questions appear to stand out. First, how much evolutionary or other unanticipated effect would be expected from genetic therapy addressed to single genetic loci (e.g., β-thalassemia or sickle cell anemia)? Second, beyond such gene therapy performed on somatic cells, how much additional benefit might be gained by reduction or elimination of such defective genes from germinal cells? Third, beyond reduction of frequency of defective genes, how much further benefit would be gained by seeking increase of favorable (e.g., disease-resistant) genes in the gene pool? These are reasonable technical questions and approximate estimates probably could be formulated.

It is to be noted that the latter two questions relate to changes of uncertain magnitude in the gene pool, but changes that in themselves are likely to be regarded as favorable. They might, therefore, be characterized as eugenic. However, the motivation of such intervention would be to improve the health of individuals and to reduce the frequency of ill health in the present population and its immediate progeny. Any alteration in the total gene pool would be likely to be quantitatively minor and a secondary rather than an intended consequence. Certainly, such secondary consequences should be assessed but they should not be confused with eugenic measures deliberately intended to produce "improved" phenotypes, particularly characteristics (e.g., "intelligence" or physiognomy) that have a far more complex genetic background than single gene defects. Such "positive" eugenic goals as creating a master race or a subservient caste are speculative in the current state of technical knowledge. We cannot be sure whether or not accomplishment of such objectives ever will be realistic.

These are questions that future generations may have to face in practical terms but nonetheless this generation cannot entirely avoid them. We have already noted that even the conjecture creates anxiety about any genetic intervention, especially among those least sophisticated biologically. To avoid what may be called a "reverse slippery slope" (a progressive inhibition of beneficial genetic and developmental research and intervention), it is desirable now explicitly to address the issues involved. It should help to enunciate policy principles as a social contract or pact among those who carry out the research or provide services and members of society who may be justifiably anxious about how far the technology should go and how decisions will be made. There should be a designed *tradition* to govern access to the human genome, a tradition that will express our highest current aspirations but will be modifiable by experience as these aspirations encounter the future.

Principles of a Designed Tradition

What might be the content of such a tradition? This is too complex and difficult a question to find a ready panacea. One may, however, suggest openers for a deliberative agenda (see accompanying box).

How might such principles achieve the status of social guidelines and how might they be implemented? First, the initial source of the guidelines should be a deliberative body of unquestionable stature, unimpeachable integrity, and universal acceptability as a forum for mobilization of the highest wisdom. It is important to emphasize that the human gene pool is potentially shared across national, ethnic, and religious barriers. People who deliberate guidelines must be protected against pressures for premature publicity but also must ensure eventual full public access both to procedures and decision making.

Philosophers, theologians, scientists, physicians, legal authorities, men and women of good will are unlikely all to speak with one voice.

Relevant expertise must be available through individuals and professional organizations who are sufficiently involved to ensure their full cooperation in later implementation. Religious and other groups whose traditions provide normative guidance to large constituencies should be fully participant in formulating and implementing the guidelines. Implementation should be not through legislation but through dedication of conscience, professional ethics, and the force of common law. Once formulated, the principles should be articulated simply, cogently, and emotively as befits a major human charter that must be understood by all and able to be transmitted as a heritage through generations.

> ## Principles of Designed Tradition
>
> • The purpose of near-term deliberate intervention in human reproduction, including heredity and development, should be to preserve the health and welfare of individual human beings and not to alter the characteristics of the species as a whole.
>
> • Preservation of health and welfare of individuals implies conservation of existing human characteristics, except as these are limiting to self-realization and satisfaction.
>
> • Correction and elimination of obvious developmental defects that are accepted as curtailing human potential should have highest priority.
>
> • No intervention should be practiced that is intended to reduce or limit human potential, whether of individuals or of groups, regardless of assumed benefits to humanity or to particular societies.
>
> • Interventions intended to gratify individual desires (e.g., sex selection) should be undertaken only when the collective consequences are judged not to be detrimental (e.g., unbalanced sex ratios).
>
> • The enunciated principles should be negotiated among and endorsed by relevant *international* professional and political entities.

Ambitious and even idealistic though all this may seem, some such social mechanism is essential to ease the advent of the coming era of enhanced genetic and developmental intervention. The recent report "Genetic Technology: A New Frontier," prepared by the Congressional Office of Technology Assessment, closes with a chapter on Genetics and Society. It notes that "The idea that research in genetics may lead someday to the ability to direct human evolution has caused particularly strong reactions." It cites the fact that "Pope John Paul II has decried genetic engineering as running counter to natural law." On the other hand, it quotes Robert T. Francouer, identified as a Catholic philosopher, as follows:

... We have always said, often without real belief, that we were and are created by God in His own image and likeness. "Let us make man in our image, after our likeness" logically means that man is by nature a creator, like his Creator. Or at least a cocreator in a very real, awesome manner. Not mere collaborator, nor administrator, nor caretaker. By divine command we are creators. Why, then, should we be shocked today to learn that we can now or soon will be able to create the man of the future? Why should we be horrified and denounce the scientist or physician for daring to "play God"? Is it because we have forgotten the Semitic (biblical) conception of creation as God's ongoing collaboration with man? Creation is our God-given role, and our task is the ongoing creation of the yet unfinished, still evolving nature of man.

This difference within one religion underlines the unprecedented character of the issues raised by the advent of this new age of intervention. Philosophers, theologians, scientists, physicians, legal authorities, and men and women of good will are unlikely all to speak with one voice. Even as we take our first steps toward outer space, our meddlesome interventive proclivity has opened the door to our own innermost space. Traditions rooted only in the capabilities and experiences of the past cannot adequately instruct us as to what lies ahead in the new millennium. We need to evolve a future-sensitive tradition that explicitly takes into account human aspirations and the ordering thrust of all of life's reproductive progression. In this task, science and technology are among the prime actors in the theatre of creative synthesis. If there is an indomitable engine or a higher historical necessity driving toward the new age, it lies in all of us and in the collective choices we make.

Even as we take our first steps toward outer space, our meddlesome interventive proclivity has opened the door to our own innermost space.

Ethical and Institutional Aspects of Recombinant DNA Technology

*Salvador E. Luria**

I believe I qualify to speak on the subject of this report on at least two distinct historical grounds. The first is that, by an accident of scientific history, I happened to be the one who, with my associate Mary Human, discovered the first instance of restriction and modification of DNA, the basis for recombinant DNA technology. This was in 1951, and the discovery was not done by working directly on DNA. What we found was that a certain bacteriophage could or could not multiply in a certain bacterial host depending on which host it had previously infected.

This discovery was immediately generalized to other systems. It was soon clear that restriction

*Center for Cancer Research, Massachusetts Institute of Technology, 77 Massachusetts Avenue, Cambridge, MA 02193

0—8412—0750—X/83/0047$06.00/0

had to do with a specific breakage of DNA when it entered the wrong bacterium. But it took several years, a decade in fact, before anyone could isolate the restriction enzymes. These proved to be a special class of enzymes that would cut DNA specifically at certain precise sequences in their nucleotide language unless certain bases were methylated. These enzymes, therefore, could be used to generate DNA fragments that could then be joined to others as desired—the basis of recombinant DNA technology. This technology resembles, at least formally, the technology of electrical circuits, with promoters, operators, terminators, and attenuators taking the place of batteries, resistors, capacitors, transformers, and so on. As I shall point out later, despite a genetic terminology that may be alien to some, the recombinant DNA circuitry is basically as simple as electrical circuitry. A good electrician is just as easy or as hard to come by as a good DNA recombinationist.

... the recombinant DNA circuitry is basically as simple as electrical circuitry.

If my first historical claim in this area is to have been an early warrior, my second claim is to have been an early worrier. In the 1960s, long before Asilomar, and guidelines for research, and P1–P4 classification or any recombinant DNA experiments I was concerned with genetic engineering of the nuclear transplant type. I worried that the new potentialities of cell biology and genetics would raise a number of issues such that the public and the scientists would not see eye to eye. I wrote a few articles on this subject, some of which received alarmist titles from magazine editors. I suggested at that time that agencies such as the National Academy of Sciences and possibly the United Nations might do well to set up committees or task forces to advise scientists and keep the public informed of the possible applications and consequences of genetic engineering. Reassurance, I believed, would make it possible to minimize conflicts, while responsible leadership would help prevent foolish or dangerous applications.

Be that as it may, by the late 1960s the actuality of a technology based on splitting, rejoining, and transferring segments of DNA was recognized. Within a decade such technology has become not only a reality but an industry—for the time being, at least, a corporate rather than a production concern. In its growth, recombinant DNA technology had to go through an adolescent crisis, in which Asilomar might be compared to an awareness of sex, the National Institutes of Health (NIH) guidelines to an attempt at parental tutelage, and the current scene to the assumption of adult responsibility.

In today's discussion I would like to touch upon three aspects of the recombinant DNA story both past and present. These are, first, the question of public health dangers; second, the problem of decision making; and finally, the ethical impact of recombinant DNA technology on institutional life.

Potential Public Health Dangers

The question of public health dangers was bound to arise with the public, in the press, and among scientists, as the awareness of DNA experimentation increased. It was thrown into sudden relief by the Asilomar conference of 1976. Press reporters, always eager to sensationalize, were at hand to feed on what for many people (I was not there) was a serious discussion of how to proceed; for some others, it was an ego trip, or rather a super-ego trip. The idea that they were dealing with potential dynamite made some scientists feel grand; the sense that they were dealing with it as responsible citizens made them feel deliciously humble if not heroic. But, as often happens in our media-infected society, the show threatened to overshadow the reality. The absolution that some scientists were hunting for, like Lewis Carroll's snark threatened to overwhelm the hunters. The snark threatened to be a bojum.

From the very beginning it was my belief as a microbiologist and a geneticist that shifting bits of DNA from cells to plasmid and to bacteria, fashioning chimeric genomes and letting them loose into our urban sewers would involve no danger beyond what human incompetence is always apt to generate. Clearly, placing genes for botulinus toxin or for plague into *Escherichia coli* would be irresponsible, as irresponsible as it would be to handle *Clostridium botulinum* or *Pasteurella pestis* incompetently in the laboratory. But I never could buy the nebulous argument that creating "a new organism that did not yet exist in nature" was intrinsically dangerous. Each time a human being is born it is a new organism, a new genome, which may turn out to be a Newton, or an Einstein, or an Attila, or a Hitler.

The discussion of dangers, as you certainly remember, was particularly bitter in Cambridge, Mass., where at times it verged on the preposterous. The City Council, having on its hands both Harvard and the Massachusetts Institute of Technology, held public hearings, where people stood up and compared recombinant DNA technology to nuclear plants and nuclear bombs. The vision of cockroaches spreading out from a Harvard laboratory to infest the city of Cambridge with recombinant monsters was raised by distinguished scientists in words fit for science fiction.

Behind the surface layer of worry there was a hidden agenda, which deserves serious analysis and to which I shall return. But the enormity of the horrors vividly evoked made it hard to discuss soberly the other issues, at least in Cambridge. Personally, I tried in a few letters to newspapers to separate the various issues, but without much success. I was blamed by fear-mongers as well as by pro-recombinant DNA colleagues, a clear indication that the problem had moved from the technical to the political arena.

For better or for worse, the sense of danger is something that tends to decrease the longer one lives with the supposed danger. This may be unfortunate in the case of real dangers like atom bombs, but it worked out well for recombinant DNA.

This is a political problem that involves questions of power and is, therefore, not soluble except possibly in the much broader framework of who does or who should wield power. But examination within the microcosm of recombinant DNA technology may be instructive.

Problem of Decision Making

The protagonists in the confrontations of 1976–77 were not divided by opposite sets of values; rather, values conflicted with goals. The scientists wanted to go ahead with their work with a minimum of regulation and a maximum of protection. They were prepared to abide by the guidelines (although most of them believed them to be unnecessary), because the guidelines represented official absolution. In fact, it is remarkable how much the guidelines were actually observed; as remarkable as I always find it that people refrain from smoking in the non-smoker section of airplanes. People don't want trouble if they are allowed to do their work.

But scientists by-and-large failed to understand, or at least to come to grips with, the real issue posed by their critics: the issue of who makes the decisions. When the decision concerns which research should be funded and to what extent the public, at least in nonrevolutionary periods, is willing to leave the decision to

Congress and to peer reviews. Even in the days of revolt in the 1960s, the question of decision making in relation to funding and performing of basic research was raised only in a perfunctory way. But the situation is different when one goes from science to technology, for example, to recombinant DNA technology. Interestingly enough, a populace that is willing to trust to its elected representatives the right to manufacture, and maybe use, thousands of hydrogen bombs can be roused by the idea that scientists may generate noxious microbes. Proximity helps propagate the concern: Harvard and MIT in Cambridge are not only educational institutions; they are employers, and landlords, and taxpayers (or nonpayers, whatever the case may be). Politics becomes a complex concern, but at its core is the unresolved dilemma: in a society permeated by technology, who decides what is to be done and what not? Is von Neumann's famous statement valid, that in the area of technology "what *can* be done *will* be done?"

... in a society permeated by technology, who decides what is to be done and what not?

I do not claim to have an answer. I tried to deal with some of the general questions in this area in a lecture entitled "Slippery When Wet" delivered at the American Philosophical Society in 1971. As evidence of the controversiality of this subject I may mention that months later I received a note from my friend Philip Handler, saying: "I am sorry I did not hear your talk. I read your lecture and I want you to know that I completely disagree."

Given that there is no generally satisfactory answer, let me try to explore the subject pragmatically for the recombinant DNA case. In Cambridge what was done was for the City Council to appoint a committee, ranging from doctors to nurses to housewives to laborers. The committee invited testimony and visited laboratories. It did what best can be done: to learn the facts and issues and to explain them to others in their report. It recommended that the city go along with the NIH guidelines but that it retain the power of ordinance control in order to reassure the public that their interests were protected locally as well as in Washington. It allowed cooling off of tempers without stifling debate. It was a good, if not perfect, exercise in democracy.

But the problem remains: who does the deciding? Essentially, it is a problem of imperfect democracy. Our elected officers cannot be depended upon to have infinite wisdom or freedom from pressures. Power in our democracy is dual: government power and corporate power, the relation between the two being generally asymmetrical. How is the public to know whether a government decision aims at the well-being of the public or is a response to corporate pressure? I personally feel that a certain distrust in the path of technology development and application is justified and desirable, not just because of physical dangers but because of possible dislocations of the power structure. I was, fortunately, enough of an expert in the field of recombinant DNA to realize the minimal level of danger. But what do I know about miniaturization of electronic circuits, or about child psychology, to be sure that TV or video games will not hurt society as I know it and like it?

The example of Cambridge, however imperfect, suggests one possible way for the public to act: a sort of local *ad hoc* communal democracy added to the electoral one in which our confidence has unfortunately been shaken. As far as scientists are concerned, it may well be that some sensible kind of machinery may be established for debating the pros and cons of emerging technologies. This may provide a forum through which to communicate with the public, its elected representatives, as well as other corporate interests in a sounder way than through the shrill voice of commercial media.

Ethical Impact on Institutional Life

The press has contributed in recent months to bring to the fore the question of the role of scientists in the industrial application of recombinant DNA technology, which is the third and last problem I wish to discuss. For once in the history of technology, scientists have in the field of DNA technology become aware of the potential fruits of their work before or at least at the same time as corporate interests. And these fruits can be the sweet ones of medical progress—the gift of Asclepius—as well as the equally sweet ones of monetary gain—the reward of Danae. Faced with the possibility of a few geneticists and biochemists becoming millionaires and many others being able to supplement their salaries by serving as industrial consultants, some sections of the press and the public, and some old fashioned university administrators, have raised the specter of corruption.

I confess that I have little sympathy for this targeted defense of academic purity. In the first place, making money out of private companies established specifically for that purpose has been the traditional way among university engineers and I believe also university chemists. In my own institution, for example, a professor may be on the board of one or several corporations, may actually found such a corporation, may consult for one or more, the requirement being that he or she not receive a regular salary and be paid only in stocks, options, and consultant fees. Conflicts of interest with the university are of course to be avoided.

Even apart from the precedents for industrial involvement on the part of university professors, I find it ludicrous that in a society like ours, where the profit motive is not only recognized but celebrated everywhere from the cradle to the Presidency—where several presidents of the United States have taken advantage of questionable loopholes in the tax laws in order to enrich themselves; in which the public watches without batting an eye while the FBI illegally entraps with bribes a number of readily bribable legislators—I find it ludicrous, I repeat, that so much fuss be made at the chance that a few scientists may become guided by profit rather than by the purest dream of discovery. Why should molecular biologists be purer or more virginal than chemists or engineers? We are none of us angels or spartans. But I am willing to bet that among all reasonable-sized fortunes that will be made in the next 20 years, those made by university scientists will be ethically at least as clean as those of lawyers, bankers, or members of Congress.

Why should molecular biologists be purer or more virginal than chemists or engineers? We are none of us angels or spartans.

Yet, the question of university research in the area of DNA technology has some more complex aspects. One aspect, which has already been debated in faculty meetings as well as in newspaper columns, is that of corporate involvement of the universities themselves in biological research. When a university owns a spaghetti factory or a car rental business, it simply takes its corporate risks like any investor. The only special problem universities have to consider is their relations to the Internal Revenue Service. Ethical considerations apply equally to universities and to other businesses.

Trouble may come, however, if corporate research sponsored by a university is done within the university itself, by individuals who are also paid to teach and to do basic research sponsored by outside agencies. Such an arrangement is quite dangerous. The scientists working on corporate research acquire a special position because they generate industrial profit in addition to or instead of the nonprofit value of teaching and pure research. They will be perceived by their colleagues as being a special and privileged group, part of the industrial rump of the university rather than of its educational front. This kind of trouble is bound to arise not only when a university goes into business for itself (something that Harvard half-heartedly decided not to do) but also when a university accepts to do targeted research specified by a corporate donor. This inevitably results in a dual system of research appointments. It might be preferable if scientists who work on such targeted contracts, as contrasted with broadly exploratory ones, were physically or academically segregated from the regular faculty, preserving the purity of the tax-exempt ghetto.

In any case, it is important to avoid having within a university department two classes of faculty scientists, those who hold purely academic jobs and those whose applied work brings a financial reward to the institution. Such a dichotomy will inevitably give rise to internal suspicions and lead to faculty demoralization. Recall the situation described, of all places, in the second chapter of Genesis: two fellows named Abel and Cain got into trouble when the big boss started making distinctions between the value of their respective outputs. As you will recall, the resulting strife led to the first major unpleasantness in creation science.

There is a scientific aspect of recombinant DNA research that is not yet generally realized and may soon affect some of the financial and industrial characteristics of this field. Until a couple of years ago, recombinant DNA research was the cutting edge of biological science. A few great experts and pioneers stood out in the same way as the great physicists of the 1920s and '30s stood out. Those were the undisputed leaders whom governments called upon in times of crisis to produce strategic bombing patterns or atom bombs.

What is happening today in DNA technology, however, is very different. The technology that may have industrial uses has proved to be so simple and unsophisticated that any properly trained undergraduate can in a few months become an expert, or at least a proficient practitioner. At the moment there appears to be little to be discovered except possibly some refinements of technique. The astuteness is almost purely commercial: to decide which gene product will sell and then go about the series of steps needed to single out the corresponding gene and put it to work. The chance of success in any one case may be reasonably greater than zero.

What I mean is that, with few exceptions, the great experts are not specially likely to become unique leaders in the industrial application of DNA research. I might do better buying stock in a large established pharmaceutical firm just entering this field than in one of the glamour companies led by my fellow scientists. Success may depend more on the wisdom of knowing what products to go after than on the refinements of scientific technology.

What this implies is that, while substantial input of industrial capital may help accelerate the application of DNA technology to practical uses, it is not needed for the advancement of molecular biology. Unless I am grossly mistaken, major advances in understanding the organization and functioning of DNA will still be made in universities. They will be made by scientists who pursue their work with little concern for applications. They may well be stimulated by what they learn while acting as industrial consultants, but they will not gain very much if while working in the lab they worry about stock options. This has been the story in chemistry: the close, intimate relation between university chemists and the chemical industry has stimulated great advances, which, however, have come mostly from the campus rather than from the factory. Wealth, or the mirage of wealth, does not hurt scientists any more than it does poets or musicians or businesspersons. But great symphonies or poems have seldom been composed just

Wealth, or the mirage of wealth, does not hurt scientists any more than it does poets or musicians or businesspersons. But great symphonies or poems have seldom been composed just to make money.

to make money. It would be unfortunate if some of the best minds in biology were distracted by financial preoccupation from pursuing their most valuable activity.

7

Recombinant DNA: A New Dimension in Life Sciences

*Irving S. Johnson**

What was the impact of recombinant DNA research on the pharmaceutical industry? I will discuss this topic within the context of discussing recombinant DNA research as a new dimension in the life sciences.

DNA in a cell serves as a stable repository of coded genetic information that can be replicated at the time of cell division to provide information to the progeny cells, as well as serve as a pattern for protein synthesis in the cell. This genetic material can be manipulated in several ways, some of which have been practiced for many years by geneticists. One of these methods is mutation. Mutations in DNA can be spontaneous due to environmental factors and errors in DNA replication, or they can be induced in the

*Lilly Research Laboratories, Indianapolis, IN 46285

0—8412—0750—X/83/0059$06.00/0

laboratory by a variety of physical and chemical agents. Mutations can lead to a change in the structure of the product coded for by the gene in question; sometimes this change in structure is so great that the product is vastly different. Other mutations may affect the regulatory elements controlling the expression of the structural gene; these mutations may lead to increased or decreased production of gene products. One of the key points to remember about mutagenesis is that it is an essentially random technique.

The second type of genetic engineering, recombination, has also been utilized for a long time. Recombination refers to exchange of a section of DNA between two DNA molecules. Recombination of DNA fragments from different organisms can occur by the mating of two organisms, where DNA is physically transferred from one organism to another. It may also occur following the use of a technique known as protoplast fusion, where one literally strips away the outer cell wall of fungi and bacterial cells and causes the remaining protoplasts—which now have only a cell membrane enclosing the components of the cell—to fuse together. These fused protoplasts contain the DNA molecules of both parents, and exchange of sections of DNA can now occur as these cells regenerate and divide. Transformation and transduction are other natural processes whereby cells exchange DNA in nature. All of these recombination processes involve random exchange of DNA sequences, and this exchange is generally limited to membranes of a single species of organism.

The power of the third type of genetic engineering, recombination of DNA, resides in its high degree of specificity, as well as the ability it provides to splice together genes from diverse organisms—organisms that normally will not exchange DNA in nature. With this technology, it is now possible to cause cells to produce molecules they would not normally synthesize, as well as to cause more efficient production of molecules they do normally synthesize. Thus, we now find it possible to synthesize human insulin, growth hormone, or interferon in rapidly dividing bacteria, as opposed to extracting these from mammalian tissues in which they are normally produced.

Thus, we now find it possible to synthesize human insulin...

Reduction of such laboratory cloning efforts to commercial production of materials of medical importance is a formidable task. As an example, I will recount our experience with the production of human insulin by recombinant technology.

Because most syntheses for insulin have low yields, the current primary source of insulin for human use is the pancreases of animals. Porcine and bovine insulin are the most commonly used mammalian insulins. Among species, mammalian insulin varies little in amino acid composition and physical structure. (Exceptions are the rat and guinea pig, which have large structural differences.)

Production of Human Insulin

In the recombinant DNA method of insulin synthesis, the plasmids (cellular elements) that contain the genetic code for tryptophan synthetase promoter, methionine, and insulin A-chain are inserted into one *Esherichia coli* culture; a second culture contains plasmids that are identical except for substitution of the genetic code for insulin B-chain. After the separate fermentations have been carried out, cyanogen bromide effectively cleaves the plasmids at the methionine residue, releasing the chains from the fused-gene product. The resulting chains are isolated and chemically modified to form insulin.

The insulin molecule contains several bonds linking two sulfur atoms. While these disulfide bonds can form in more than one way, only one of the arrangements yields a product identical to pancreatic human insulin (PHI). Compounds that contain the same number and type of atoms, but differ in the way the atoms are joined, are called isomers. Isomers do not necessarily function identically. Insulin has two theoretically probable isomers; in producing biosynthetic human insulin (BHI), care must be taken throughout the synthetic process to guard against contamination of BHI by one of these undesirable isomers.

Minor structural differences between substances in a mixture (for example, a mixture of isomers) can be detected by using several types of sophisticated analysis. In the case of insulin, samples of the pure isomers were tested several ways; when BHI was tested, the results were compared. Identical test results for two samples indicate equivalent chemical properties for the two samples; the converse is also true. Testing indicated that the undesirable isomers were not present in detectable quantities in the synthesis product. This same comparison testing was also used to prove that the product, BHI, was chemically equivalent to PHI.

...the recombination DNA method of producing [insulin] rules out contamination with other hormones...

In contrast to the procedure for the production of animal insulins, the recombinant DNA method of producing BHI rules out contamination with other hormones, such as glucagon, somatostatin, or even proinsulin.

After chemical equivalence was shown, it was essential to show that biologic activity of BHI was also identical to that of PHI. Insulin potency can be measured by using a standard procedure, and the comparative effects of BHI, PHI, and porcine insulin were identical (1).

A variety of analyses were performed to determine if BHI, PHI, and porcine insulin exhibited

similar physical chemical properties. The identical results strengthened our hypothesis that BHI is chemically equivalent to PHI (2).

In the case of any product such as insulin, which is intended for continuous use over prolonged periods of time, there is a need to ensure not only that the material is pure, authentic, and active, but also that it is as free as possible from any adverse effects. Human insulin is a foreign protein to laboratory animals, and its main effect is to reduce blood glucose levels. The prolonged survival of animals to which human insulin is administered is difficult to achieve. BHI was subjected to acute toxicity studies in mice, rats, and dogs; to studies over 14-day periods in monkeys; and to studies over 30-day periods in rats and dogs. All these studies demonstrated evidence only of reduction in blood glucose levels, and not of any other adverse effects.

Many protein materials used in medicine may be contaminated with pyrogens, that is, substances which raise the body temperature. Two tests for pyrogens have been used in the study of BHI. A number of BHI lots were tested, and none of the lots proved to be pyrogenic (3).

Another test relating to pyrogenicity is used to detect what is sometimes called endogenous human pyrogen, although the precise nature of this pyrogen is unknown. The method of detection involves incubation of human peripheral leukocytes with the test preparation. After incubation and centrifugation, a rabbit pyrogen assay is done on the supernatant liquid. Muramyl dipeptide is an inducer of endogenous human pyrogen and is used to show that the assay is capable of responding. Porcine insulin alone is nonpyrogenic but becomes pyrogenic if combined with the inducer. Proinsulin and BHI, whether made by the proinsulin route or by the combination of the A- and B-chains, are noninducers.

The procedures . . . open up new possibilities for future investigation in the field of treatment of diabetes.

Theoretically, a possible source of protein contamination of BHI could be derived from the *E. coli* organisms used in its manufacture. A complex assay method for detection of such contamination was developed, and all lots of BHI are subjected to the assay. Patients receiving BHI have been tested for specific antibodies, which could result from protein contamination, and results show that after 6 or 12 months of treatment the antibodies are not present.

An alternate route to BHI is via human proinsulin, obtained by inserting the proinsulin plasmids into *E. coli* K12. The end product is treated to give crude proinsulin, which is chemically modified to yield insulin. BHI made

64

via proinsulin is essentially identical to BHI made via chain combination (as previously discussed).

BHI obtained from proinsulin was subjected to a battery of tests; additional tests were used to determine the levels of C-peptide, proinsulin, and cleavage enzymes in the final insulin product.

The presence of human proinsulin in human insulin may be advantageous. The human pancreas secretes proinsulin along with insulin in the ratio of about 1:100. Possibly, mixtures of insulin and proinsulin, or insulin and C-peptide, or even all three, may ultimately be used in diabetic therapy. Human proinsulin is, therefore, not only of interest as a route to BHI, but also for the possibilities it offers of novel therapeutic approaches.

The procedures developed for the manufacture of BHI by the proinsulin route open up new possibilities for future investigation in the field of treatment of diabetes. Many of these possibilities are now being actively explored by scientists and clinicians throughout the world.

The exhaustive studies developed demonstrate the essential identity of BHI and PHI, and verify the extreme degree of purity and freedom from undesirable contaminants required for a drug intended for human use. Existing testing methods are continuously improved as the technologies involved grow.

Cytokines

Immunology, to my mind, is the second cutting edge on a most exciting area in biology. Moreover, recombinant DNA and modern immunology overlap and complement each other. The term immunology is frequently associated with vaccine production. Recombinant DNA will no doubt lead to improved vaccine production by fermentation. However, I would like to focus on the general class of immunologically important proteins known collectively as *cytokines.*

Cytokines are protein molecules that are made in one type of cell as a result of a stimulus and leave that cell to affect the function of a second type of cell. Although the most well-known examples of cytokines are the proteins called interferons, there are more than 50 cytokines that have been identified on the basis of their biological activities. Many of these cytokines may later turn out to be identical polypeptides obtained from different cellular sources and characterized biologically by different workers. More definite structural characterization, which by and large is not available except for one or two of these important molecules, is required before this identification ambiguity can be resolved. One of the problems associated with cytokines has been the very low natural tissue concentration, which has hampered the determination of their biochemical characteristics. It is interesting that the complete structures of the various interferons were established only after the genes were cloned by recombinant DNA technology. This sequence may prove to be the case for many of the other cytokines as well.

Cytokines can affect inflammatory responses, enhance certain types of cellular responses (a type of helper function), and may serve as suppressors. These types of activities have potential application in a wide variety of clinical problems.

Production of Antibiotics

The other area of application of recombinant DNA in the pharmaceutical arena that I would like to discuss is that of antibiotics. From the standpoint of market size—some $6 billion worldwide—antibiotics represent one of the largest potential industrial targets for recombinant DNA in the immediate future. Antibiotics differ from all of the previous cases because they represent not primary gene products but rather so-called secondary metabolic products. In other words, they are synthesized in cells by reactions catalyzed by enzymatic proteins that are primary gene products. Most antibiotic biosynthetic pathways involve a number of steps involving the enzymatic conversion of a series of metabolites to other substances.

...antibiotics represent one of the largest potential industrial targets for recombinant DNA...

Recombinant DNA technology can be applied to antibiotic production in at least three ways: yield improvement, *in vivo* modification, and hybrid antibiotics.

Any one of these various enzymatic conversions could be rate-limiting due to limiting amounts of the enzyme necessary to catalyze the reaction. By careful analysis of the accumulation of intermediates in a set of reactions, it is possible to determine the rate-limiting steps, and then to clone the gene coding for the particular enzyme catalyzing those steps. If multiple copies of that gene can then be introduced into the antibiotic-producing organism, enzyme levels should increase. The end result would be the removal of the synthetic rate limitations.

Today most antibiotics used medically are chemical modifications of the actual biosynthetic products, for example, the large group of cephalosoporins and penicillins available for clinical use. Many of the chemical reactions carried out on a natural antibiotic to improve its level of activity or broaden its spectrum of activity can also be carried out by using enzymes. Usually these enzymes do not, however, occur naturally in the producing organisms. It is now possible, using recombinant DNA, to introduce gene coding for these enzymes into production strains. It is also possible to accumulate large amounts of intermediates (ones that nature might never make) for use in chemical modification.

The idea of hybrid antibiotics comes from the concept that novel structures may result from adding genes for enzymes related to the biosynthesis of one type of antibiotic to an organism producing an entirely different type of antibiotic. We can do such an experiment in a defined way (more like in vivo modification), or we can simply "shotgun" genes from one producer into a cell producing a second antibiotic and screen for novel activities. In a sense, the latter process is similar to the exchange of genetic information that has occurred between organisms throughout history at a very low frequency and that has been part of the evolutionary process. These shotgun combinations could be looked on as a way of enormously speeding up evolution.

Recombinant DNA and Agriculture

I would like to turn now from the realm of pharmaceuticals to that of agriculture. Agriculture is, without doubt, the world's largest business. In the United States alone, the assets of this business are on the order of $900 billion. Agriculture employs some 15 million people in the United States, and it accounted for about $40 billion in exports in 1980. Agriculture naturally occupies a central role in the maintenance of human life. Yet, as we look to the future, it becomes very clear—based on current population growth rates—that by the turn of the century our present conventional agriculture will be in serious difficulty. We probably won't be able to feed the world's population without some radical improvements. Some applications for recombinant DNA technology that were developed for pharmaceuticals can be adapted to the animal husbandry area.

One of the most exciting areas in agriculture, however, is the modification of plants. Increased crop yield can be achieved by increased acreage, by increased yield per acre, and by a combination of the two. In any case, genetic improvement in crop species will be important. Much of the available uncultivated land in developed countries, such as the United States, is marginal for crop use. To use that land effectively, large quantities of fertilizer and frequent irrigation may be necessary. The cost of nitrogen fertilizer continues to escalate because of increases in petroleum costs; continued irrigation eventually leads to severe damage to the soil because of salt accumulation as the salt-containing water evaporates from the soil. Most currently used crop species have a very low tolerance for the salt in their soil, and soil irrigated for many years will simply no longer support growth. A classic example was the irrigated soil of the Euphrates Valley of biblical times; the salt content of the soil became so high that plants withered and died, as did the culture they supported. Salt accumulation in the irrigated soil of the San Joaquin Valley in California results in large annual losses in crop production. In the Imperial Valley in California some crops have actually gone out of production because of soil salinity.

We probably won't be able to feed the world's population without some radical improvements.

To increase crop yield per acre, we must find ways of increasing the efficiency of energy utilization in plants as well as ways of causing the plant to devote more of this energy to production of edible parts. Until recently, plant breeders have used mainly classical genetic techniques, not unlike those used by Gregor Mendel before the turn of this century. Two of the most serious impediments to an efficient plant-breeding program are the amount of time that is required between cycles of effective selection and the difficulty of maintaining sufficient genetic diversity to allow for selection of rate gene combinations. Today a number of new genetic techniques promise hope for the future. The development of methodology to maintain individual plant cells in tissue culture, coupled with the technology for regenerating whole plants from such an individual cell, has led to the beginning of a revolution in plant breeding. It is possible today to carry out, in weeks, a breeding step that would previously have taken an entire growing season. As a logical extension of tissue culture technology, the technique of protoplast fusion allows us to strip away the cell walls of two different plant cells, fuse them together into one cell, and then regenerate a plant from the fused cell. The possibilities for moving important genes from one plant species to another and thereby increasing the genetic diversity of crop plant cells are enormous.

Salt and drought tolerance are actually very closely related. Whenever the salt concentration of the water surrounding plant cells is increased, cellular dehydration can result in death of the plant. The same holds true when water availability is affected by drought. Some organisms (including plants, animals, and microorganisms) have evolved mechanisms for combating dehydration. These mechanisms may involve taking salt into the cell from the environment to balance the salt concentration inside and outside of the cell, or may involve the biosynthesis of organic molecules inside the cell to prevent dehydration. Here, the goal of genetic engineering is to identify the genes controlling these mechanisms in salt-tolerant species and move those genes into important agronomic crops.

These applications are limited only by the imagination and desire of the researchers, by resources, and by new information.

Salt tolerance breeding programs are underway throughout the world for several crops including alfalfa, barley, corn, grapes, and tomatoes; and some of these programs involve various aspects of the new genetic engineering technology. Several different genes will no doubt be involved, and the interaction of these genes in specific ways may be important for the net beneficial result. Obviously, it is much more difficult to move several genes from one cell to another and to regulate the function of these genes in the new cell than it is to move a single gene, such as that for insulin, into a cell where it may even be allowed to function essentially unregulated as long as the protein product is produced.

Improvements in photosynthetic efficiency and in the development of plants whose cells can fix nitrogen directly from the air could lead to re-markable improvements in crop yields. Nitrogen fixation is a very energy-intensive process. If the enzyme systems for nitrogen fixation were operative in plant cells, the energy supplies of these cells could be taxed severely. Thus, unless a concomitant increase in photosynthetic energy production by the plant is possible, the positive effect of nitrogen fixation may be negated. A number of scientists are working to improve the efficiency of photosynthesis, and some promising leads have been identified. Genetic engineering—recombinant DNA, in particular—will play an important role in these experiments.

One area that may prove practical in the near future is improvement of the nutritional value of certain plant proteins. If the genes for these proteins can be cloned, it may be possible to increase the amount of certain essential amino acids in the protein by reconstructing part of the gene. The technology is in hand to do this. The difficulties begin when attempts are made to return this engineered gene to the plant. Further developments in genetic technology are needed to accomplish reinsertion successfully.

In Summation

I have presented an overview of possible applications of genetic engineering in the pharmaceutical and agriculture areas. These applications are limited only by the imagination and desire of the researchers, by resources, and by new information.

I would like to leave you, however, with the thought that perhaps one of the most important uses of recombinant DNA will not necessarily be to produce new products directly. Recombinant DNA can play a role in the elucidation of the fundamental mechanisms used by cells of all types to perpetuate their existence. All phe-

All phenomena of life relate eventually back to the structure and regulation of genetic information.

nomena of life relate eventually back to the structure and regulation of genetic information. The orderly process of differentiation of cells into specialized tissues, as well as the seemingly uncontrolled growth of malignant tumors, is ultimately determined by genetic information. As we understand these processes better, we will find chemicals, produced in traditional and nontraditional ways, that can be used to influence the fundamental process of cellular growth and function.

Literature Cited

1. Chance R. E.; Kroeff, E. P.; Hoffmann, J. A.; Fank, B. H. *Diabetes Care* **1981**, *4*, 147–154.
2. Chance, R. E.; Hoffmann, J. A.; Kroeff, E. P.; Johnson, M. G.; Schirmer, E. W.; Bromer, W. W.; Ross, M. J.; Wetzel, R. In "Peptides: Synthesis, Structure and Function. Proceedings of the Seventh American Peptide Symposium"; Pierce Chemical Company: Rockford, IL, 1981; pp. 721–728.
3. Ross, J. W.; Baker, R. S.; Hooker, C. S.; Johnson, I. S.; Schmidtke, J. R.; Smith, W. C. In "Hormone Drugs: Proceedings of the FDA–USP Workshop on Drugs and Reference Standards for Insulins, Somatotropins, and Thyroid-Axis Hormones"; United States Pharmacopeial Convention, Inc., Rockville. MD, in press.

Science and Business: Working Together to Meet Human Needs

*Robert A. Swanson**

Some people these days seem to think that there is something a little suspect in any working relationship between the research scientist and the businessperson. Such an attitude displays a limited awareness of the history of science. For there have always been such relationships, and humankind has benefited from them. As Louis Pasteur once said, "Science and the applications of science (are) bound together as the fruit to the tree which bears it."

One example in physics illustrates that this relationship is a two-way street. James Watt's invention of the steam engine led to the enunciation of the laws of thermodynamics. Rudolph Diesel, in turn, employed those laws of thermo-

*Genentech, Inc., 460 Point San Bruno Boulevard, South San Francisco, CA 94080

0—8412—0750—X/83/0073$06.00/0

dynamics to invent the diesel engine. Even Pasteur's discovery of optical isomers resulted from his work with the wine industry.

My remarks will focus on the symbiotic relationship between science and business, using recombinant DNA technology and the experience of my own company, Genentech, Inc., to illustrate the points I want to make.

A two-way flow of information and creativity is taking place today in our laboratories at Genentech. We have brought together a wide range of scientific disciplines and a diverse number of observations from the basic scientific literature to produce needed products. In turn, solving the problems involved in making these products a reality has stimulated new discoveries and provided key insights to some basic research questions. We are proud of this interaction and feel that it is one of our Company's key strengths.

Others have recognized the importance of this kind of interaction. Alfred North Whitehead, the great mathematician and philosopher, said that one of the greatest discoveries of the 19th century was "how to set about bridging the gap between the scientific idea and the ultimate product."

Although there is nothing new about the relationship of science and business, today the nature of that relationship is in a rapid state of change.

Economic Changes Dictate Research Funding Changes

Part of the change is economic. Prior to World War II, the business community provided a principal source of financial support for university research, both basic and applied. During the war, the federal government began underwriting the costs of research on a very large scale, and after the war the government's support of scientific research kept increasing over the years.

Today that government support is being cut back sharply. The cutbacks are the inevitable result of a fiscal crisis that must be resolved by the federal government. The driving force behind them is quite independent from the enormous recognized good that support of scientific training and research has provided our country.

But, economics is only part of the reason why everything is in a state of change today. Overlaying the recognition of government limits is the unfolding of molecular biology as a science, and the application of one of its most important tools, recombinant DNA, by industry. This is not a small change and has been described even in *Science* magazine as "a revolution in Biology." It is a change that will allow new approaches and leaps forward in a broad range of scientific disciplines. There will be major shifts of people as skill needs change, and industry will provide new job opportunities and a shift in the sources of research funding.

The impact on university science is great. Not only are universities rethinking their traditional relationship with industry, they are also faced with the more complex problems of relationships with the new biotechnology companies, many of them started by their own professors. Already the questions are extensive.

How much time should faculty be permitted to devote to outside business interests?

What kinds of relationships are permissible between university faculty and private business?

How do universities share in the rewards of commercialization when they made contributions to the basic science?

Should a university invest in its own business spin-offs?

There are no easy answers to these and other questions that are now being addressed on university campuses. Each institution will of course work out its own solution. But the one thread woven through all the solutions will be a greater interaction between science and business. One of the most important aspects of this interaction is the creation of new high technology companies. It is a process that is a uniquely American phenomenon, created in part by our frontier spirit, and must be carefully protected because it is critical to the long-term vigor of our economy.

In this regard, the recombinant DNA industry is fulfilling the historic mission of new, innovative business. Like prior new high technology companies started in the field of computers and semiconductors, the recombinant DNA industry is in the process of making American industry more competitive in world markets at a time when we as a nation desperately need to be competitive. It is producing jobs when the nation urgently needs employment opportunities. It is at the threshold of delivering products that will enrich people's lives.

... the recombinant DNA industry is in the process of making American industry more competitive in world markets....

In 1967, a Commerce Department study found that more than half of all U.S. inventions and innovations were accounted for by small business and individual inventors. In 1976, a Massachusetts Institute of Technology Development Foundation study found that young technology companies far exceeded their larger, more established competitors in the rates of sales growth, taxes paid, and especially the number of jobs created. In the past 10 years, according to a new report, small, innovative businesses have created 3 million jobs, while net employment in the 1,000 largest U.S. corporations has remained more or less level.

The case of my own company, Genentech, Inc., illustrates many of the points I have been making. Our company is just 6 years old, yet already three products of our research are undergoing the clinical testing that is required before marketing approval can be given: human insulin for the treatment of diabetes, human growth hormone for the treatment of growth problems in children, and interferon, which shows promise because of its antiviral and antitumor effects.

All of these products have been made by designing microorganisms using a technology Genentech helped to pioneer. These products would not be on their way to economical availability today without the commercialization of recombinant DNA technology.

During this 6-year period, we have created over 325 new jobs and spent close to $35 million in research and development. We have also invested over $20 million in facilities for research, manufacturing, and administration that now exceed 144,000 square feet.

However, we have yet to sell a single ounce of product to an end user. The enormous quantities of money required to finance our business during the product development period, as well as during the time necessary for the recognized and understandable regulatory approvals, have come from private risk capital, a public stock offering, and the licensing of a portion of our technology. Although we have a long way to go to achieve our goal of building a profitable, billion-dollar company, our start-up phase will be completed this year with the anticipated market introduction of our first product.

The Genentech story began in the early '70s.

Development of Genentech

My partner in starting Genentech, Dr. Herbert Boyer of the University of California at San Francisco, had been working for over 15 years on a number of scientifically interesting but publicly obscure basic research projects, funded in part by the government. In 1973, some of his observations led him to collaborate with Dr. Stanley Cohen of Stanford University, who had been working in a complementary field. Together they demonstrated that it was possible to transfer foreign DNA into a microorganism and to have it become part of that organism's genetic structure.

The news of their success was received with great interest throughout the scientific community and elsewhere, but several years would pass before the commercial implication of their work would be widely recognized. When I first contacted Dr. Boyer in early 1976, we were among the very few who believed that useful products could be developed around this technology in a reasonably short time frame. As the British scientific journal *Nature* later said, in 1975 "even optimists would have predicted that it would be a decade before genetic engineering would be commercially exploited." I guess Herb Boyer and I were more than optimistic, if there is such a thing. Within 6 months of our meeting, we had formed Genentech, developed a business plan, and persuaded a group of venture capitalists to risk their own money on our ability to prove two things: that the technology was ready for commercialization, and that we could build a business.

In the following year, 1977, we at Genentech expressed the brain hormone somatostatin, which was the first useful protein to be produced by recombinant DNA technology. The late Dr. Philip Handler, then president of the National Academy of Sciences, described our achievement as a "scientific triumph of the first order."

In the 4 years since then, we have produced many more products by this new technology. Besides human insulin, human growth hormone, and leukocyte interferon, which I mentioned previously, Genentech has produced fibroblast and immune interferon; calcitonin, which is important in the treatment of some bone diseases; albumin, which is used to replace lost blood; and our first products for agriculture—bovine and porcine growth hormones, and a vaccine for foot-and-mouth disease developed in collaboration with the U.S. Department of Agriculture.

Our research laboratories have many other new products, pharmaceutical, agricultural, and industrial chemical, at earlier stages of development.

In an article discussing Genentech's success in producing immune interferon, the British scientific journal *Nature* said recently:

> *The latest achievement by Goeddel and his colleagues gives testimony to the tremendous accomplishments now possible with recombinant DNA technology when multiple resources are integrated into an attack on a specific problem. If one considers the many methodologies and types of expertise required to accomplish this work, it is clear that it could only have been achieved in a large industrial research center.*

In other words, not only can top quality science be conducted in an industrial setting, but solutions to some of today's complex scientific problems can benefit when the focused multidisciplined approach of industry is wed with the creativity and excellence of academic science. The relationship between science and business clearly benefits science as well as business.

To reach the full potential of biotechnology, we need to focus on the challenges posed by all innovation—challenges involving patent protection, regulatory affairs, scientific and technical education, and funding for research. In these important areas, discussions will help bring industrial and academic science closer, and government and university policy decisions will critically impact the success of a fragile new industry. Here again, I think that Genentech's experience underlines the importance of these concerns.

Role of Patent System

Innovation cannot exist without a strong patent system. Contributions by new businesses to our national economy are made possible, to a considerable degree, by the protection given to innovation by patents. Without the potential for protecting our developments, Genentech would not have been able to raise the substantial capital needed for our growth and to sustain us during the period while our products go through the regulatory approval process.

If the products of our invention were to become available too soon to others who had not incurred the same research and development cost, the risk confronting a would-be investor would be too great. After all, what farmer will invest in seed if the law permitted others to take his crops?

Patent protection also makes for a healthier economy by strengthening competition. Under the umbrella of a patent, a new company can compete against larger, older, and more entrenched corporations. This, in turn, stimulates the older companies to increase their own research and development efforts, and to lower prices on older products now challenged in the market place by new products. Both of those results are desirable, I believe, from a public policy point of view.

With respect to regulation, the government should continue to analyze the costs as well as the benefits. The impact of regulation often falls most heavily on new businesses, where an infrastructure for handling the paperwork is not yet developed and where the high costs are spread mainly over a few newly introduced products. Even before we have sales, at least 5% of Genentech's employees are dealing primarily with regulatory matters.

Responsible business can and does regulate itself. A good example of the benefits that can flow from a realistic, common sense, and flexible approach to regulatory issues took place at the beginning of our industry. In 1976, the National Institutes of Health established a Recombinant

DNA Advisory Committee, usually referred to as RAC. RAC set up guidelines governing recombinant DNA experiments. Those guidelines were mandatory for institutions and persons receiving federal funds, but all companies working in the field also volunteered to abide by them. By late in 1981, after reviewing the record carefully, RAC came to the conclusion that some of its requirements could be relaxed because the safety of the new technology had been well established. No federal regulations were ever enacted.

This approach—vigilant and demanding, but flexible and constructive—has enabled recombinant DNA technology to make progress in this country at a rate unparalleled anywhere else in the world. Japan, for example, where regulation of this new technology has been rigid and restrictive, is several years behind the United States in this field.

The Food and Drug Administration (FDA), too, has taken a constructive attitude in making the benefits of this technology quickly and safely available to the public. In our own experience, the FDA has not required unnecessary studies

Federal Regulation

Responsible business can and does regulate itself.

81

and has provided us with technical assistance and encouragement as we began the process of taking our first product through the approval system.

In our case, the FDA has been demonstrating that regulatory burdens can be minimized and that urgently needed drugs can be speeded on their way to patients, without in any respect lowering the agency's high standards for proof of safety and efficacy.

Need for Scientific and Technical Education

The third challenge that I listed is scientific and technical education. Thanks to the genius of American scientists, the courage of our venture capitalists, and the cooperative attitude of the NIH and the FDA, the United States now enjoys a significant lead in the application of recombinant DNA technology among the nations of the world. But this should not give anyone a feeling of complacency, for that lead is threatened, not by failures in American inventiveness, but by potential shortages in personnel. Highly skilled personnel are vital to this industry—indeed, to any high-technology industry. At Genentech, one of every five employees has a PhD degree. Although we are having no difficulty in attracting the highly qualified men and women we *now* need (thanks in large part to our firm commitment to quality science and peer recognition, including encouraging publication of research papers), government, industry, and the universities will have to act quickly if a bottleneck to growth, of the kind other high-technology industries have already experienced, is to be avoided.

As a nation, we must encourage our universities to train more scientists and engineers.

Scientifically and technically trained men and women are a critical resource for keeping U.S. industries at the forefront of technological progress. In this respect we are clearly falling behind. For example, in 1979—the most recent year for which figures are available—the United

States graduated approximately 16,000 electrical engineers, while Japan graduated 22,000. In other words, Japan, with about half the population of our country, graduated nearly 35% more electrical engineers than we did. One result of this trend has been the recent inroad Japanese companies have made into the semiconductor industry founded by our country. I have only given the figures for electrical engineering graduates, but I am afraid the picture is similar in other scientific and technical disciplines.

As a nation, we must encourage our universities to train more scientists and engineers. But that brings up the next challenge I mentioned. It is also important that we provide adequate funding for basic research, that great pool of technical information from which future inventions can be drawn. After all, Dr. Boyer's early work was part of that pool, and at the time no one would have predicted how it could eventually be applied.

Despite the cutbacks in federal funding that I spoke of earlier, I am sure that the government will continue to provide support for such research. Without it we will destroy our seed corn. On the other hand, a greater part of the financial support for university research can and should come from the private sector.

There are many ways in which the universities can acquire needed funds through arrangements with business, *without* impinging on academic freedom. Here, too, patents can play a part. For example, Stanford University and the University of California share the Boyer-Cohen patent on DNA recombination. I understand that so far more than 70 companies (Genentech included) have each agreed to make substantial royalty payments to the universities when they practice under that patent. In other cases, government funding of basic research can be supplemented by industry grants for graduate training, consulting agreements, and research contracts. Approximately 5% of Genentech's own research expenditures are spent as part of university interactions. The impact would be great if our larger cousins did the same, targeting their support to the training of future scientists and engineers.

University researchers in the last 30 years have laid the groundwork for the rapid rise of the biotechnology industry. Academic biologists and biochemists collaborated little with the business sector as they made their landmark discoveries. This is now changing. While all transitions are stressful, this is a healthy development with substantial benefits to be harvested by both academia and industry. Perhaps most important, a close working relationship between science and business, between the university and the corporation, will ensure that the fruits of this technology will quickly reach the public.

Management of Recombinant DNA Research

Philip H. Abelson*

This publication treats many aspects of a field that is very important for research and development and for service to the needs of society. Recombinant DNA is one of several techniques that will be applied in large-scale creation of food, pharmaceuticals, agricultural chemicals, and major chemical products. Ultimately, biotechnology will be involved in making items having an annual value of the order of $100 billion or more. Recombinant DNA is also having and will continue to have a place in fundamental research as a means of addressing many central questions of biology, such as: What are the mechanisms involved in gene expression? What causes some genes to be turned on in some tissues while other genes are turned on in other tissues?

*Science, 1515 Massachusetts Avenue, N.W., Washington, DC 20005

0—8412—0750—X/83/0085$06.00/0

Discovering and implementing the potentials of biotechnology will entail a very large research activity, the development of industrial processes, and the creation of production facilities, followed by marketing efforts. Where will the many research efforts be conducted? To what extent and in what roles will university scientists be active? What part will the newly created biotechnology companies play? What roles are the large pharmaceutical and chemical companies likely to have? What will be the relationships between academic and industrial scientists and between universities and industry? To what extent will traditional free scientific interchange be curtailed or hampered by efforts to obtain patents and other proprietary advantage? What roles will the federal government have?

Discovering and implementing the potentials of biotechnology will entail a very large research activity....

I will not attempt to give detailed answers to all these questions. Many of the relationships are changing and they will continue to do so. But I will outline some of the factors that will shape the outcome and then make some guesses about the future.

Role of Universities

I visualize a secure role for the universities. They are the prime site of education, the source of trained people for all the activities of research, development, and production. In addition, the universities have been the home of fundamental research in biology. If we include medical schools, the universities have been the home also of biochemistry and other sciences relevant to biotechnology. This has been particularly true of the body of knowledge leading to recombinant DNA. It was academically centered studies that created a broad base of knowledge about synthetic mechanisms in *Escherichia coli*, the behavior of plasmids, the genetic code, and the existence of restriction enzymes.

The principal motivation for these studies was the search for knowledge and understanding. Most of the work was performed before the slightest possibility of practical applications could be envisioned. From the 1950s to the early 1970s, little or no work relevant to recombinant DNA was performed by industry. In fact, during this period activities in industrial microbiology centered on the production of pharmaceuticals, including antibiotics, and the production of a few chemicals such as citric acid and some amino acids. Before 1950 there was substantial production of ethyl alcohol, acetic acid, acetone, and 1-butanol, but this activity stopped when petrochemicals synthesized from cheap oil flooded the market. A 10-fold increase in the price of oil has changed the economics of the situation, a factor that I will discuss in more detail later.

Results of the crucial experiments that established the possibilities of recombinant DNA were revealed to a scientific audience in 1973. Those present quickly recognized the importance of the findings. They were excited by the new vistas that they could imagine, but they were also fearful of possible injurious effects that uncontrolled experimentation might bring. Soon afterwards the Asilomar Conference was organized and this led to prudent precautions. Initially, the concern expressed by scientists may have dampened enthusiasm for large-scale use of recombinant DNA. However, research continued. A crucial experiment was performed by Struhl and coworkers and reported in May 1976. Their work showed that a eukaryotic gene, for example, one from yeast, could be expressed by the prokaryote *Escherichia coli*. This appeared to indicate that a gene from any organism could be inserted into *E. coli* and might be successfully cloned.

Big industry was slow to recognize the potentials of recombinant DNA. Pontecorvo in 1976 berated a meeting of industrial microbial geneticists for their failure to exploit the new technique. Industry appears to have viewed molecular biology as useless. However, in 1976 two scientists thought otherwise. They scraped up a thousand dollars of their own money to form a fledgling company. In this activity and in their behavior, these scientists followed a path much like that physicists had followed 50 years earlier. A quotation from a recent article by Michael Jacobs in the January *Physics Today* illuminates the situation in molecular biology at this point:

> *The phenomenal growth of physics as a science over the past half century has been paralleled, breakthrough for breakthrough, by the explosion of physics as a marketable commodity. Technology has followed close on the heels of research, and has sometimes even taken the lead and acted as a stimulus for research. As often as not, those responsible for bringing the latest technologies to the marketplace have been the physicists themselves, too devoted to the technologies to trust their development to others, unable to interest any existing industries in the unproven innovation, or just too shrewd to let the valuable technologies slip through their fingers. If anything unites this group of physicist-businessmen, it is their unerring dedication to their technologies, and the dream of establishing them as items of utility and profit.*

One of the more successful of the new firms formed to exploit recombinant DNA is Genentech. Its founders included Robert Swanson and Herbert Boyer. Boyer was one of the inventors of the recombinant DNA technique. Genentech in

1977 pioneered in the production of somato-statin, a brain hormone. A synthetic gene was inserted into a plasmid of *E. coli* and the bacteria's synthetic apparatus was used to produce the hormone. In 1978 they produced insulin and in 1979 human growth hormone. These were very important achievements scientifically and attractive commercially. The total market for insulin is about $400 million a year. A more dramatic event was the announcement in June 1980 of the synthesis of interferon with recombinant DNA. Interferons are widely regarded as potentially extremely valuable substances. At present, they are our best hope for successful treatment of viral diseases. Earlier experiments had demonstrated the efficacy of interferon in combating viruses. More recent research has raised hopes that it might be successful in treating cancer. Were all these hopes to be realized and successful products obtained, their value might be of the order of several billions of dollars. Because of the potential medical benefits of the interferons, great efforts had been made to produce some to test their effects. Before the production announced by Genentech, only tiny amounts had been produced by tissue culture methods. The cost for a few milligrams was millions of dollars or, to state the matter with more impact, $22 billion per pound. Talk of a product with a potential market in the billions of dollars has the capacity to elicit excitement. Two months after its announcement of the synthesis of interferon, Genentech made a public offering of 1.1 million shares of stock. About 7,500,000 shares are outstanding. The stock was sold at $35 a share. On paper, Boyer, who owns 925,000 shares, was worth more than $30 million. The offering also produced a day of excitement on Wall Street. The stock that was sold at $35 quickly rose to $89 a share, later to fall back somewhat.

The spectacle of such financial fireworks had profound effects on academic molecular biologists. It also impressed executives in industry who had previously not been much interested in recombinant DNA.

Academic scientists rushed to form companies. Approximately 150 have been incorporated. Today most of the academic experts, near experts, and not so experts in the field have some kind of corporate connection. Some of them are on the staffs of new companies. Some have been hired by larger established corporations. Many professors remain at universities while serving on the boards of directors of companies and holding an equity position in the companies.

The example of one of their colleagues suddenly becoming wealthy was, of course, a powerful factor in causing molecular biologists to go after money. But there was an additional stimulus. The conditions under which research is conducted at universities have deteriorated. Time spent in preparing grant proposals has become excessive. Bureaucracies at the universities have increased, as have the overhead charges to support them. Government regulations and the demands of government auditors have become annoying and onerous. The ties of loyalty to the universities have been weakened.

University presidents viewing the contemporary scene have their own set of problems. Private universities have been hit with cutbacks in student aid and burdened with federal regulations imposed during the past decade. Many public universities have financial problems due to actions of state legislatures. Thus, leading research universities face financial problems and uncertainties about federal support of research. Some of them have indicated a desire to obtain part of the riches that seem to be flowing from recombinant DNA. At the same time, however, they find that key staff members are making their own arrangements under terms that are relatively unprecedented. Research universities have long been accustomed to arrangements under which faculty members consulted with industry. In general, no more than 1 day per week was involved and in most cases the frequency was much less. Such arrangements were viewed as benefiting all involved, including the university. The faculty members who consulted learned of problems that interested industry and were in a position to place their students well. The activity improved university–industry relationships and often led to some financial support by industry of research and fellowships at the universities.

The new situation in which faculty members have equity positions in companies cannot be regarded by university presidents as being as healthy as the earlier consultancy arrangements. At any rate, questions have arisen. Will communication among scientists be attenuated as each attempts to discover or acquire information of potential proprietary value? Will professors use the work of graduate students to enrich themselves and outsiders? Will tensions arise in departments as some professors garner money while others do not? Can a professor simultaneously serve two masters? Faithfully and well? The presidents of leading research universities are concerned about these questions and unsure about what their policies should be. Some are getting together to mull over these matters.

Prospects for Recombinant DNA Companies

What is the nature of the new companies that are causing such financial waves and concerns on the part of university administrators? For the most part, they are small in size but large in intellect. The excitement of molecular biology and its fruitfulness as a field of research, coupled with good federal support, earlier had drawn to molecular biology the best young minds. The new companies have captured many of these top-flight people.

Detailed statistics are not available on the capital assets of these new companies or on the total number of their personnel. They range in size from a few organizations having assets of about $100 million to many whose capital is only a few thousand dollars. A few of the companies have about 70 full-time PhD scientists, others have no full-time staff. For reasons that I will set forth, it is my guess that not more than five of the 150 companies, and perhaps only three or four, will be viable 5 years from now. However, those that survive will make important contributions.

The costs of development, production plants, and marketing far exceed those of research.

Two factors that will impinge heavily on the conduct of the new small companies are the costs and time delays that intervene between a research result and a profit-making product. The costs of development, production plants, and marketing far exceed those of research. A commonly accepted formula is that for every dollar spent on research it is necessary to spend $10 on development and $100 on production facilities and marketing.

Ordinarily, the time span between research and a profitable product is 4 years or more. In the case of biologicals for human use, the delays can be extremely long. Costs and delays involved in obtaining clearance from the Food and Drug Administration are such as to create difficult barriers for the introduction of new pharmaceuticals. This is especially true for small companies.

A significant factor in clearance for human use is the need to avoid toxic substances in the product. *Escherichia coli*, which has been the favorite organism for recombinant DNA research, produces an endotoxin; isolating a product that is free of toxicity will require special care. The regulatory people can be counted on to be suspicious of products formed by using *E. coli*. Fortunately, other organisms such as *Bacillus subtilis* or yeasts can also be employed in recombinant DNA synthesis. These organisms do not produce substances toxic to humans. However, their genetics are not as well known as that of *E. coli*. Microbiologists who have limited their biological experience to one organism may find it necessary to broaden their capabilities.

The problems of FDA clearance that I have just outlined have not dismayed Genentech or their partners, Eli Lilly, in efforts to produce insulin with *E. coli*. Insulin is not a new drug. It has been used for about 50 years. In the process invented by Genentech, the *E. coli* produce large quantities of an insoluble moiety that contains the insulin. This solid can be readily freed from the bacterial toxins. Lilly, a highly reputable pharmaceutical company, is constructing a plant to produce insulin by the Genentech process. Clearance problems are not expected.

It appears that the situation is similar for human growth hormone. This substance, too, has been produced by other methods and has been employed clinically.

... when new products are synthesized, the delays and costs of clinical testing and FDA clearance will be large.

However, when brand new products are synthesized, the delays and costs of clinical testing and FDA clearance will be large. As the tiny new companies seek to identify new lines to work on, they will find that most of the relatively easy to make and potentially profitable products have already been produced or are under study by larger groups. The largest of the new biotechnology companies have better capital resources, more scientific and engineering competence, and a head start over the small latecomers.

Four new venture companies that have gained respect and admiration among scientists are Genentech, Cetus, Biogen, and Genex. Many regard Genentech as the leader. In addition to somatostatin, human insulin, human growth hormone, and interferons, it has produced thymosin alpha-1, a dozen subtypes of interferon, bovine and porcine growth hormones, calcitonin, a vaccine for hoof and mouth disease, and human albumin. No company, large or small, in the pharmaceutical business can boast of such an outstanding record of achievement in recent times. The company also has an unusually liberal policy about publication of results. Once production of a given item is successful, detailed information is made available in the peer review journals. *Science* has published a number of the research reports. Genentech, located in San Francisco, has a staff numbering 350, of whom 70 have a PhD degree. Last year the company spent about $17 million on research and development. It produced a tiny profit. It will be some years before substantial profits are forthcoming. However, the many products already in the pipeline guarantee substantive revenues at a later date.

Cetus, headquartered in Berkeley, Calif., is the one major recombinant DNA company whose activities antedate the discovery of recombinant DNA. The company was formed in 1971. It has produced some of the same products as Genentech, but its program of developments has been broader. Cetus uses every important tool of genetic engineering, including recombinant DNA, cell fusion with protists, monoclonal antibodies, immobilized cells, and enzymes. It has been successful in engineering better microorganisms for fermentation. Cetus has excellent intellectual resources and excellent relations with academic scientists. It has assets of about $125 million and it showed a small profit in 1981.

Biogen has laboratories in Cambridge, Mass., and Geneva, Switzerland. To a degree, its products parallel those of Genentech. It has produced interferons, which are now in the testing phase. In late February 1982, Biogen announced successful trials against viruses of common colds. Major thrusts are devoted to vaccine, including those for hoof and mouth disease and hepatitis. It has activities in bacterial engineering and fermentations. Biogen has a total staff of 150 people, including 50 PhDs, and capital of $65 million. The company is privately owned. Its backers include International Nickel, Schering–Plough, and Monsanto. One of its assets is participation by Walter Gilbert, a Nobel Laureate.

Genex has two laboratories near Washington, D.C. Its activities include research on recombinant DNA. It also engages in various facets of bioengineering, including technology assessment and studies of scale-up economics. Its first product that has been sold is aspartic acid, which is used in making the artificial sweetener aspartame. It has produced interferon for Bristol-Myers and urokinase, which prevents blood clots in the lungs, for a Japanese firm. It has also made human serum albumin. One of its efforts is devoted to the production of alcohol with thermophilic organisms. Genex states that it has dozens of projects under way. It believes that its strength lies in a staff of high-powered engineers who work closely with biochemists and geneticists. The company has 220 employees, 55 of whom have a PhD degree. Genex is privately owned. Its participants include Monsanto, Emerson Electric, and Koppers Company.

From this brief analysis of the leading four venture companies, it should be evident that small competitors are up against a tough situation. A few may survive through the brilliance of their staff, concentration on limited objectives, and some luck. And in spite of the enormous resources of some of the nation's industrial giants, many of them may also encounter stiff competition from such companies as Genentech.

An important factor determining the possible roles of academic molecular biologists will be

Small Businesses Versus Established Firms

the activities of various major pharmaceutical, chemical, and petrochemical companies. These companies have begun to move to ensure their participation at the frontiers of biotechnology. As I have already indicated, early developments in recombinant DNA caught most companies flat-footed. They had not sponsored research in the field, their scientists were not experts in it, and they were basically incapable of judging the significance of the new developments. When the academic scientists formed companies to exploit commercial applications of recombinant DNA, large companies were among the providers of venture capital.

... *early developments in recombinant DNA caught most companies flat-footed.*

A sample of major companies who have obtained an equity interest in the new ventures includes Allied Corporation, American Cyanamid, Standard Oil of California, Dow Chemical, Eli Lilly, Monsanto, Phillips Petroleum, Rohm and Haas, and Standard Oil of Indiana. The typical investment is on the order of $5 million. For this, the major companies obtain a window on the trends and potentialities of recombinant DNA research. Some obtain patents or a preferred position with respect to patents that might be forthcoming. For the companies, the relationship has been a comparatively cheap method of getting educated and obtaining expert evaluations and advice. However, they are free to establish their own in-house capabilities and they are under no obligation to put additional money into the new companies. Until the new ventures make a profitable product, which may be many years hence, they will experience a hemorrhage of capital. Unless they can develop exciting products of high potential market value, they will find that their erstwhile benefactors will channel funds elsewhere. In view of the costs of engineering, construction of plants, and marketing, it is unlikely that more than a very few of the new companies will actually market a product. Rather, their best hopes are to develop useful organisms and processes that can be patented and to obtain royalties from the patents.

In mentioning organizations that have equity positions in recombinant DNA ventures, I named three that are major petroleum companies, along with large chemical concerns. Why should oil and chemical companies be interested? I believe one reason is that they wish to continue to exist in the 21st century. Another reason is that they too are on the lookout for new, highly profitable products.

A brief history of petrochemicals may provide a useful background. Before 1930, there was little use of petroleum as a feedstock. It was only after 1950 that this use was more fully exploited. At that time and for more than two decades after, oil was by far the best and most versatile cheap feedstock. It was exceeded for some applications only by natural gas (or methane), which was extremely cheap. The major oil and chemical companies came to base their production almost entirely on petroleum and natural gas. For example, Dupont obtains 85% of its products from these hydrocarbons. Annual production of chemicals from petrochemical feedstocks has totaled more than $100 billion. All the major oil and chemical companies are engaged in the production of petrochemicals. In the course of three decades, research led to many products and to great competition among the manufacturers of these items. The easy-to-discover, high value added products were identified. What remained was grueling competition for shares of the markets.

The events of the 1970s convinced people that supplies of oil are limited and exhaustible. The 10-fold increase in the price of oil and a much larger increase in the price of natural gas changed the competitive position of these substances as feedstocks. What will be the feedstocks of the future? For countries that have it, coal will be important. A substantial fraction of chemicals will be derived from synthesis gas, that is, carbon monoxide plus hydrogen. Rivaling coal in many applications will be renewable feedstocks based on photosynthesis. The pre-

Petrochemicals: Recombinant DNA Opens New Doors

Why should oil and chemical companies be interested?

97

eminent substance will be glucose. This sugar will be obtained in part from plants but probably mainly from the cellulose of trees. Today, vegetation typically fixes less than 1% of the solar energy falling on it. There is very substantial room for improvement. Already through selection of superior stock, yields of wood have been improved greatly. When modern forest management, all the techniques of tissue culture, and recombinant DNA are harnessed, a very large amount of wood will be available as a feedstock.

Glucose is a favorite metabolite of many microorganisms. They can convert it into many hundreds of compounds. Already some of these fermentation products can be obtained from glucose more cheaply than they can be made from petroleum. For example, industrial alcohol (ethanol) can be produced more cheaply by fermentation than through petrochemistry. The competitive edge will probably be improved through a combination of biotechnologies that include recombinant DNA.

Fermentation processes can be used to obtain many chemicals that are now made from oil. The annual sales of such substances runs into tens of billions of dollars. It is not necessary to create a market for the fermentation products—the market exists. It is only necessary to make the products more cheaply than they can be obtained from oil or coal. For the long term, a major commercial opportunity lies in finding profitable means of converting wood into useful products. In this effort, biotechnology will have an important role. Even before the potentialities of recombinant DNA became obvious, some of the major petroleum and chemical companies were establishing positions in biological chemistry. They were interested in many parts of the food system, both plants and animals. Part of the motivation was undoubtedly the recognition that a growing population will require more and more food. Producing that increased amount of food will require better seeds, fertilizers, herbicides, pesticides, growth hormones, and nutrition supplements such as amino acids. Here again is an existing market that is virtually guaranteed to be an increasing one. Many companies already report that their agrichemical business is one of their most profitable lines. Its annual magnitude is tens of billions of dollars.

Food and Medicinals

For the biotechnologist, one of the major challenges is to improve the yields of crops. An obvious goal which has been repeatedly identified is to create a corn that can fix nitrogen. This goal may prove elusive. It may also be difficult for experts in recombinant DNA to quickly establish a role in the improvement of seeds for crops. There are already experts in that field who have been and continue to be successful in the improvement of crops through applied genetics. They are entrenched in the seed companies. They understand the requirements of farmers, which vary from region to region. Their progress is slow but it is relatively certain. For the experts in recombinant DNA to have an impact they must succeed in creating varieties that are substantially better than existing stock. In efforts to improve seeds, biotechnologists will have available relatively new techniques of plant tissue culture and protoplast fusion which facilitate introduction of genes into cells.

The meat industry is a potential market for many biologicals. An example is animal growth hormones. Genentech has already created a suitable recombinant DNA process to produce the hormones for testing by Monsanto.

In the field of medicinal chemistry, recombinant DNA has interesting potentials in addition to that of interferon. Many proteins and peptides are involved in the regulation of body functions. Examples are the endorphins, which are naturally occurring opiates. My guess is that as human biochemistry and neurochemistry are better understood, many additional peptides and proteins will be found to influence health and behavior. Recombinant DNA techniques should be particularly effective in the production of such substances.

An alternative approach to treatment of viral diseases is to prepare vaccines specific to each of them. Merck Sharp & Dohme, the pharmaceutical company, has been successful in preparing such a vaccine against hepatitis B. It uses recombinant DNA techniques to produce an antigen, which, when injected, stimulates production of antibodies in humans.

Move to In-House Capabilities

Of all the large companies DuPont was the first to move with respect to acquiring an in-house capacity to utilize recombinant DNA techniques. This move began in 1979. It was a natural extension of DuPont's then existing participation in biotechnology. For DuPont, life-science-based sales were $1.5 billion in 1980. In 1982 DuPont will spend $190 million on research and development in life sciences and will spend an additional $100 million on new research facilities for the life sciences. The research will, among other things, involve use of recombinant DNA and monoclonal antibodies. The company believes that engineered cells will enable the production of a wide variety of products. Of greatest interest to DuPont are those related to health care, agricultural applications, and chemical feedstocks. They plan to use fermentation microbiology to amplify the use of biological systems to make products.

Ralph Hardy of DuPont stated recently, "It appears reasonable that there will be a restoration of fermentation for production of selected feedstocks, especially oxychemicals." Hardy also stated:

Several key steps are required to develop a successful process for a given chemical feedstock—recombinant DNA may be one of the steps. Adequate quantities of a low cost fermentation substrate must be identified. Organisms with the genetic/enzymatic capabilities must be found. These capabilities are manipulated by a variety of methods so as to equal the established goals. The desired genes may be collected into a preferred host. Other critical steps include fermenter design and an economic, energy conservative, product recovery. The competitiveness of these biological systems with conventional chemical systems will depend on the total package—biological and engineering—and their timing will vary with the specific feedstock.

Another facet of DuPont's planned activities is agriculture, and again I quote Hardy:

Examples of potential products are seeds with agrichemical, disease, or insect resistance. Chemical seed combinations where a seed is genetically modified so as to be highly resistant to a potent broad spectrum herbicide is an example ... Additional opportunities include resistance to temperature and water stress, improved quality of the harvested part of the plant, and increased yield, which is the major opportunity.

I have mentioned some of DuPont's activities because they are an example that other major companies are already beginning to follow. In a year or two, we may expect that as many as 20 companies, each with assets of $1 billion or more, will be conducting internal biotechnology research and development employing recombinant DNA. These large companies are able to pay high salaries to recruit many scientists. They also can afford to make substantial investments in research and development. Some of them are willing to make long-term commitments to important research goals.

I have touched on some of the factors and forces that will shape where and how recombinant DNA research is conducted. I have indicated my belief that much of the applied aspect will be conducted in large companies. Some of it will be performed in the limited number of new companies that survive. Most of the small organizations will disappear. However, both the government and industry will support research at universities. Demand for trained people will continue and probably grow. The kind of research that individual professors undertake will reflect the source of support. The really basic work will be done under the auspices of the government. However, ties to industry will not be all bad. With time, the various faculties will evolve standards of ethics and behavior that will make it feasible for professors to have ties with industry.

The excitement of interacting with experts joined together in a common venture and the incentive of shared rewards are powerful stimuli for overachievement.

Universities will not be quite the same 5 years hence as they were 5 years ago. Their financial pressures have induced a search for additional sources of support. There will be greater emphasis on applied research on campus. The major research universities now have patent policies, and greater efforts will be made to obtain royalties. A sizable fraction of the university faculty, one that previously had little contact with industrial research, has now joined engineers, chemists, and physicists in search for practical applications.

The major new venture companies like Genentech, Cetus, Biogen, and Genex will have a struggle to survive and prosper. But one or more are likely to emerge as major companies. What they lack in financial capital is counterbalanced by their intellectual assets. They have many of the best people, and they can retain them by furnishing them with an equity position in the company. The excitement of interacting with experts joined together in a common venture and the incentive of shared rewards are powerful stimuli for overachievement.

Having been caught fast asleep on recombinant DNA, leading research companies can be expected to try to avoid similar mistakes. They will increase their ties and interactions with the universities. They will also develop their own strong biotechnology capability.

10

Regulation of Recombinant DNA Research and Industry

George E. Brown, Jr. *

The initial promise of recombinant DNA technology has already resulted in a few products reaching the marketplace. These few, however, are undoubtedly the initial trickle of a flood of products sure to follow.

You may already know the history of the 1977–1978 legislative foray into regulation of recombinant DNA research. Briefly, bills were reported by Committees in the House and the Senate that would have regulated recombinant DNA research, but they were never debated or put to a vote in either House, so they died at the end of the Congress. The Subcommittee on Science, Research, and Technology, under then Chairman Ray Thornton, did its part by holding 12 days of hearings on the issue and then consid-

*Representative from California, 2342 Rayburn House Office Building, Washington, DC 20515

0—8412—0750—X/83/0105$06.00/0

My hope is that the
example ... of reasoned
and rational examination
of recombinant DNA
research will be followed
by other legislative
committees.

ering the bill. Chairman Thornton, after due consideration, came to the conclusion that the bill should not pass.

My hope is that the example set by the House Science Committee of reasoned and rational examination of recombinant DNA research will be followed by other legislative committees having jurisdiction over various aspects of this new field.

I believe all of us can learn the following lessons from this now historic episode in science policy making.

Relationship of Congress and Scientific Community

First, whoever dreamed up the legislative process was a genius to the extent that the process is slow enough to prevent a stampede of unwise legislation, and, second, Congress, in my judgment, had and continues to have, an extremely healthy opinion of the scientific community. Congress is willing to listen to well-reasoned arguments and, believe it or not, many Congressional Representatives actually have the capacity to change their minds when confronted with new and rational arguments.

In this case, Congress proceeded with fairly good judgment in advance of any realized hazard. The educational process undertaken by the Science Committee was, I think, critical in the derivation and implementation of a good, working set of guidelines for research in this area. Over time, as our knowledge increased and our anxiety over the risks posed by the research decreased, these guidelines have undergone an evolutionary downgrading such that now more than 90% of recombinant DNA experiments are exempt.

Recently, a proposal to rather drastically change the heretofore mandatory guidelines to a voluntary code of standard practice was introduced in the NIH Recombinant DNA Advisory Committee.

The Chairman and two Subcommittee Chairmen of the House Science Committee expressed their concern over this potentially far-reaching development in a letter to former Congressman Ray Thornton, who is now the Chairman of the NIH Recombinant Advisory Committee. So, although legislative interest in this area is at a low level currently, general interest by members of Congress is not. The outcome of the most recent Recombinant Advisory Committee (RAC) meeting again illustrated the wisdom of this deliberative body. Members of the Committee voiced their concern that a perceived diminishing of the federal role in this area might lead to an avalanche of conflicting state and local regulations that could seriously hamper the progress of recombinant DNA research as well as the development of the industry. As one RAC member put it, "We will lose the opportunity to move the technology along with uniformity." The result of this discussion was that the guidelines will remain mandatory but will continue their previous evolutionary trend in a scientifically rational manner.

Role of Government Policy

I would like to outline a few of the coming challenges that I see, particularly in light of the burgeoning expansion of companies entering the biotechnology arena, from the perspective of someone involved in government policy making.

As products and processes developed through biotechnology begin to enter the marketplace in large numbers, regulatory agencies will become increasingly involved. The growth of the industry itself will draw more federal legislators as well as state and local officials into direct involvement. Already, the National Institute of Occupational Safety and Health has introduced a proposal for worker health and safety in production plants utilizing recombinant DNA technology, and several municipalities have moved to regulate genetic engineering research and development located in their communities.

The patent system has not yet been deluged by a flood of applications that will force it to develop a system that avoids the delay of case-by-case litigation. The legal system has not yet had to face the possible question of establishment of limits on proprietary rights conferred by patents on the products and processes involving biotechnology. Moreover, the questions and problems of ensuring proprietary rights internationally have not yet been fully dealt with. But these problems are likely in the future.

The federal government can, and has, I believe, encouraged the commercialization of biotechnology by a program of enlightened participation and, to some extent, anticipation.

There is, surely, no lack of investor interest in recombinant DNA applications, thus venture capital is plentiful. By giving those individuals wishing to invest in biotechnology (or other high technology industries) appropriate incentives, the federal government can further encourage the growth of innovative and productive industries. In addition, legislation currently before the House (and already enacted in the Senate) would allow small businesses to apply for federal grant money to do scientific research (whether it be in genetic engineering or some other field). If applied in an enlightened manner, such legislation might work to speed up the time between an important scientific discovery and its potential application to human needs and could do this without disrupting the funding base for the universities, the nurseries for both industrial and academic research scientists.

Another potential area of concern is the worldwide competitive position of the United States in biotechnology. Much time and effort has been spent in recent years analyzing the decline of the U.S. competitive position in other technologies such as steel, automobiles, and consumer electronics. These analyses have shown that U.S. inventions are often used by our competitors to beat U.S. companies in the marketplace. This has led some people to conclude that we should limit the flow of technological information. However, in science there is a long and distinguished history of openness and exchange of information. This has been particularly true in the life sciences. I believe it would be a mistake to sacrifice this scientific freedom for the sake of, perhaps, only a short-term competitive advantage. Companies based overseas have already moved to invest substantial amounts in the biotechnology efforts of the United States. To try to restrict technology transfer at this stage would be akin to shutting the barn door after the proverbial horse has escaped and would smack of protectionist barriers to free trade. Certainly, the scientific community here and elsewhere would bridle at the thought of not being able to discuss theories and experiments with their foreign counterparts.

In this connection, I should mention that the Office of Technology Assessment, an arm of the U.S. Congress, has undertaken to follow its very popular completed work "Impacts of Applied Genetics" with a study that will attempt to analyze both the health of the American biotechnology industry and the competitive position of the United States in this area. The House Science Committee was one of the committees requesting that such a study be done.

The patent laws of many other nations, however, are far stricter than those of the United States in that no open discussion of a potentially patentable product or process may occur until the patent is published. This fact may work to inhibit scientific exchange and, as I mentioned previously, foreign proprietary rights of U.S. patents may be jeopardized.

Concerns have been voiced recently about the

changing relationship between the universities, industries, and the government. In the biotechnology field particularly, many companies are either formed by or have offered substantial equity to academic research scientists. Such arrangements raise questions as to their propriety and possible conflicts of interest. In addition, substantial industrial investments in selected university programs with concomitant exclusive patent rights have concerned several legislators who fear that corporations may be skimming the cream from years of federal investment in basic research. Indeed, two subcommittees of the House Science and Technology Committee held hearings on these concerns last June. Although certainly not frivolous, such concerns are, perhaps, overstated. Nevertheless, several meetings between university presidents and their counterparts in the biotechnology industry will be held in the coming year to attempt to set some ground rules for faculty participation in industry.

One must keep in mind that there is a continuing lack of public understanding about many aspects of modern science and technology, including biotechnology, and by this I mean not only recombinant DNA, but other equally promising technologies such as hybridomas and immobilized enzymes. Admittedly, the attention of the public and press has been focused on genetic engineering, specifically its possible *hazards*, and this attention has decreased over the last few years. However, the decreased level of attention is *not* due to the attainment of a better understanding, but rather to a lack of crises or "Andromeda strain" events. Any such event, a near-accident, a real accident, an investment fraud scheme, could easily revive public interest, only on a larger scale commensurate with the expanded biotechnology operations. A public clamor for action could force Congress to make the kind of hurried decisions on policies and regulations for which it is famous (or infamous) in many industrial sectors. The last time, an ordered, rational examination of the issues undertaken by the Science and Technology Committee succeeded in stopping some unwise legislation. At some future time, we might not be so fortunate, and we could find that misinformation has become law.

In discussing the industrial applications of genetic engineering, we must move into a larger arena where we will encounter a much larger group of people and a host of new problems affecting technological advancement. The small group of skilled research scientists who never before needed to worry about patents, or the Occupational Safety and Health Administration, or the Environmental Protection Agency, or the Food and Drug Administration, or the Federal Trade Commission has moved from benchtop experiments to pilot plant production, from blackboard theory to 10,000-liter fermenters. Genetic engineering has followed chemistry and nuclear physics into the often calamitous world of commerce.

Genetic engineering has followed chemistry and nuclear physics into the often calamitous world of commerce.

Self-Policing in Genetic Engineering Industry

To avoid overly restrictive regulations being forced on it, the genetic engineering industry might be wise to follow the lead of the research community and make an effort to establish a self-policing system. The NIH Recombinant DNA Advisory Committee has functioned extraordinarily well in this regard and has earned the respect of scientists and nonscientists alike. An industrial biotechnology association has been formed recently, and this might serve as a nucleus around which public discussion, minus proprietary details, could coalesce.

Such a discussion could focus on health and safety concerns, legal questions, patenting, and licensing as well as product registration and application processes far better than a cumbersome governmental inquiry. Whatever group might undertake such a task would undoubtedly consider it burdensome, but it would certainly be less burdensome than a regulatory apparatus forced upon the industry in the heat of public passion.

The NIH Recombinant DNA Advisory Committee has established public trust in the rationality of those involved in genetic engineering through its open actions and inclusion of public members. An industrial association or council could accomplish the same purpose for the next phase, commercialization of the technology. The inclusion of public members, especially ones who could evaluate the ethical and moral questions sure to arise, will probably cause objections, but experience has shown these individuals to be valuable contributors to a sometimes esoteric discussion.

People are generally not as fearful of something they understand.

Perhaps the most important function of such a group would be educational. Efforts could be made to familiarize legislators, regulatory agencies, and state and local governments with important developments. Perhaps, more importantly, such a group could serve an educational function within a small public forum, say a member's constituency, or in society at large. People are generally not as fearful of something they understand. Communication between industry, research scientists, and the lay public in advance of any dispute or crisis can foster understanding and thus lessen a draconian public reaction to crises that might arise.

Establishment of such an industrial group would certainly serve to further the impression of this industry as a responsible and concerned sector of our economy.

I am optimistic about the future of science and technology and, in particular, about the biotechnology revolution. One only has to read the recent scientific literature in the life sciences to learn of hosts of valuable new discoveries; and one only has to scan the written and broadcast media to hear of the potential applications of those discoveries to pressing human problems: diagnosis and treatment of life-threatening diseases, better and safer vaccines, improved crop yields. I don't think I am being a Pollyanna when I say that biological science has entered a time of unprecedented discovery zeroing in on the very essence of life.

... biological science has entered a time of unprecedented zeroing in on the very essence of life.

The establishment of an industry centered on this revolution is, I believe, an historic event. It poses a unique challenge to the best minds in research, law, and government to anticipate the industry's future direction and to involve and educate all elements of society who have an interest in this area. Such farsighted planning is difficult and has rarely taken place. Witness the current problems of the nuclear and toxic chemical industries.

A broad-based planning effort, encompassing anticipatory self-regulation and public education such as might be carried out in the biotechnology industry has never been accomplished. Other technologies have depended upon the public's unwaivering and often unquestioning support of the new and innovative. Toxic waste dumps, oil spills, and nuclear accidents have changed that. Science for too many today involves risks publicized and benefits taken for granted.

If this industry, or even academic research scientists, follow the road that appears easier initially, the cloistered avoidance of other societal forces, the penalty to be paid years hence may be great. A small investment in openness today will ensure the continued growth and prosperity of an industry that holds great promise for humankind. Indeed, the technology itself will advance in research laboratories across the country, but if public backlash against a heedless industry makes it impossible to bring the technology to the marketplace, it will remain as merely interesting scientific reading.

11

Summing Up: Science and Society in Transition

*Earl D. Hanson**

Our goal was to examine some of the ways in which recombinant DNA research affects our lives. The views presented are diverse, coming from scientists involved with recombinant DNA research and those somewhat distant from it, either in terms of their scientific discipline or because of a decision to put recombinant DNA research in another perspective. Also present were industrialists and a legislator and they saw things from yet other perspectives. Even with that range of input and also, because of it, there are certain obvious omissions evident in this gathering of ideas.

Little is said about bioethics and its spectrum of concerns: human experimentation and its many concerns for researchers, for institutional

*Science in Society Program, Wesleyan University, Middletown, CT 06457

0—8412—0750—X/83/0115$06.00/0

... what are the proprietary rights of those who create life forms?

review boards, and for those experimented upon; assignment of responsibilities, legally and morally, if ever there is an actual worst-case scenario; and so on. Related to the foregoing are philosophical issues concerned with the nature of life as we must sense it now as also a humanly contrived product and the revisions that can entail in both religious and secular perspectives.

Also, legally, what are the proprietary rights of those who create life forms? What is the public accessibility to those forms? These points are touched on in the context of patent rights. And that also points to a shortcoming of this book, which is the limited treatment accorded certain topics that are mentioned.

The first point to be made, then, in this overview is simply a recognition of certain real limitations on what is covered, both in terms of breadth and depth. That tells us that one objective was to underscore the great range and complexity of the issues raised by this research, especially when we consider its impact on our lives. And in that lesson we come to the very positive achievements of these presentations for they do contain significant thoughts, both as explicit themes as well as some implicit insights.

Potential Social Benefits

... greater yet will be the added understanding of the nature of life as seen from this powerful genetic viewpoint.

Clearly, the most insistent message was the extraordinary promise of this research. Great as are the potential social benefits, greater yet will be the added understanding of the nature of life as seen from this powerful genetic viewpoint. It is a refrain sounded by Potter and Kiefer in their introductory remarks, restated in yet another way by Luria and Abelson, and receiving yet another emphasis by the industrialists, Swanson and Johnson. And then Grobstein's sobering concerns on how to educate ourselves to the advent of our roles as genuine creators of life are echoed in the practical concerns of Brown, the legislator, who rightfully emphasizes the need to regulate only enough to assure public confidence in and continued research by scientists on these extraordinary opportunities to intellectually inform and physically benefit humankind.

To fully enjoy these opportunities we will need new social systems. These are presently evolving as Luria, Abelson, and Swanson point out. But the size, organization, and especially the modes of interaction between present social institutions are not at all clear. The cooperation between academia and industry is considered inevitable by all and potentially beneficial by all but with various qualifications needed to safeguard those benefits, for it is a process which must surely change the thinking of those in academia and industry as adjustments are made. Adjustments there must be in educational programs to meet industry's personnel needs; adjustments are needed, also, in industry to ensure the flow of new scientific ideas but without intruding on or distorting that flow; adjustments must occur in individual thinking and values so as to assure full service to one's employer—university or corporation—but also to share one's professional competence in responsible ways with others who need that professional knowledge.

The cooperation between academia and industry is considered inevitable ...

Regulation and Responsibility

There is still the unresolved question as to just what form regulation of recombinant DNA research should optimally take. Other countries, notably England and Japan, have governmental regulation. But Brown points out that the legislative process in the United States, combined with the voluntary role of DNA researchers and the explicit formulation of sensible National Institutes of Health guidelines, has achieved what was needed. Not without some turmoil as Kiefer, Grobstein, and Luria remind us, but, overall, there was no undue harassment of scientists nor lapses of responsibility toward the public welfare. Luria derives the lesson from the conduct of affairs by the Cambridge City Council that "local *ad hoc* communal democracy" is one viable answer to this vexsome question of who regulates what. But, along with others, he still emphasizes the deep complexity of issues here and the need for continuing monitoring of this crucial arena.

Perhaps the social institutions that will evolve furthest and fastest are the presently embryonic (though with multimillion dollars in assets) biotechnology corporations. Abelson predicts that of the roughly 150 in existence now, perhaps only five will be functioning 5 years from now. Will they be subsidiaries of larger corporations? Who will be their stockholders? And even more important, what will be their established image? Benefactors of society, we all hope, but will the image be tarnished by an excess of commercialism or bright with a sense of public service? Will these companies be known as accessible repositories and sources of new ideas or as competitively secretive?

Our Role as Creators

Lastly there are two special perspectives that emerged here. One is Grobstein's awareness of the creative role that is thrust upon us with the advent of the new interventionist potentials of recombinant DNA research. As he rightly asserts, we have been creators all along but, until now, more as slow manipulators of forces we have patiently come to understand in life around us, as by selective breeding. Now we choose what to make, in terms of chemical hereditary information, and insert it into a living system to function there as a new genetic endowment. The enormity of that new power has not yet really sunk in. He is surely right in urging a thoughtful, penetrating, and responsible exploration of those vistas as soon as possible. Coupled with that, there must be equally cogent proposals as to the best modes of disseminating and, if necessary, regulating (that problem again!) the consequences of those deliberations.

The other perspective, only mentioned tangentially by a few speakers is the image of science that emerges when one considers recombinant DNA research and its impact on the human prospect. This is a time when the general image of science is widely admitted to be diminished from that of a decade or two ago; when support of science is under especially skeptical review; when the ever-expanding horizons of scientific knowledge must also include vistas blighted through the thoughtless use of powerful, scientifically-based technologies and where powerful institutions, ostensibly serving science and technology, are too often mindless of individual dignity and rights. Scenarios have been written that speak of recombinant DNA research as contributing to such scenes. Less extreme but nonetheless cautionary is Brown's remark that "Science for too many today involves risks publicized and benefits taken for granted."

In the evolution that lies ahead, involving the further institutionalization of recombinant DNA research, the contingent biotechnologies and the way they are perceived in the public mind will have much to say about how the public comes to view what science is. We are still in an age where the public image of science is shaped largely by the uses of atomic energy. In that context, Hiroshima and Three Mile Island loom larger than the innumerable peaceful uses of radioactive isotopes. If indeed we are entering an era of biotechnologies—and who can seriously doubt it—it is that area that will do much to inform and condition attitudes and public policies that will determine the role science will and can play in our lives. The human prospect that is affected by recombinant DNA research includes the societal context which regulates the health of science itself.

Selected Additional Reading

Abelson, J., "A Revolution in Biology," *Science,* 1980, Volume 209, pp. 1319–1321. (This article is the first one in an issue devoted to recombinant DNA. All articles are written for experts and offer well-documented insights into the exciting state, 2 years ago now, of research in this area.)

Cavalieri, L., "The Double-Edged Helix. Science in the Real World"; Columbia University Press, New York, 1981. (This book reviews the interactions of science and society as they relate to the pursuit of recombinant DNA research.)

Dyson, F., "Disturbing the Universe"; Harper and Row, New York, 1979. (This contains a short discussion on p. 169 of the dangers of playing God through recombinant DNA techniques.)

Goldoftas, B., "Recombinant DNA: The Ups and Downs of Regulation," *Technology Review,* May/June 1982, Volume 85, pp. 29–30, 32. (A recent review of what has happened regarding the complex issue of controlling recombinant DNA research.)

Grobstein, C., "The Double Image of the Double Helix"; W. H. Freeman, San Francisco, 1979. (One of the best books on this subject with appendices that reprint some of the key documents.)

Jackson, D. A.; Stich, S., Eds., "The Recombinant DNA Debate"; Prentice-Hall, Englewood Cliffs, N.J., 1979. (A source for examining many of the bioethical problems that surround this debate.)

Krimsky, S., "Genetic Alchemy: The Social History of the Recombinant DNA Controversy"; MIT Press, Cambridge, Mass., 1982. (A forthcoming volume that tells the recombinant DNA story in a social context.)

Olby, R., "The Path to the Double Helix"; University of Washington Press, Seattle, Wash., 1974. (Probably the most detailed history to date of the various sciences that contributed to our present understanding of DNA.)

Office of Technology Assessment, "Impacts of Applied Genetics. Micro-organisms, Plants, and Animals"; Public Communications Office, Office of Technology Assessment, U.S. Congress, Washington, D.C., April 1981. (A balanced, clearly written report summarizing a wide range of applications of a variety of genetic techniques, including recombinant DNA, to social needs and including risks, patenting, and public involvement in decision making.)

Watson, J. D., "The Double Helix. A Personal Account of the Discovery of the Structure of DNA"; Norton Critical Editions in the History of Ideas, G. S. Stent, Ed., W. W. Norton and Co., New York, 1980. (This edition contains, in addition to Watson's original text, various commentaries and reviews, which add greatly to Watson's justly famous story.)

Glossary

of Selected Terms

Bacteriophage	A virus whose host cell is a bacterium.
Biomass	The amount of biological material, living or dead, in a system.
Endotoxin	A poison produced within a cell and usually only released when the cell disintegrates.
Enzyme	A protein molecule that catalyzes chemical reactions; it regulates the rate of reaction.
Epigenetics	Development of a fertilized egg into an adult, especially when viewed as a patterned unfolding of developmental events.
Eugenics	The improvement of humanity by selectively altering its genetic composition.
Eukaryote	A cell with nuclei bounded by a membrane and containing double-ended chromosomes that undergo mitosis and meiosis. (Also spelled eucaryote.)
Gene	A Mendelian factor located on a chromosome; a definable locus on a chromosome; a mutable hereditary unit; a hereditary unit that determines the formation of a specific protein; a specifiable sequence of **nucleotide** bases in a molecule of deoxyribonucleic acid (DNA). (The last two definitions are the ones most widely used. In total, these definitions indicate the history of the gene concept and its progressive refinement.)
Genetic code	The **nucleotide** triplets in DNA that, through the intermediation of messenger RNA, specify the amino acids of a given protein.

Genome	The total genetic information in the cell of an individual.
Genotype	The genetic constitution of an organism.
Heterochronic hybridization	Bringing together in one cell the genes of species that arose in different evolutionary times, e.g., the insertion of a human **gene** into a bacterium.
Interferon	A protein produced by vertebrate cells in response to viral infection.
Microbe	A general term for a unicellular organism, usually applied to bacteria.
Molecular genetics	The study of heredity in chemical terms.
Nucleotide	A purine or pyrimidine molecule attached to a ribose or deoxyribose sugar that, in turn, is attached to phosphoric acid. The nucleotides found in DNA are adenine and guanine as purines, and cytosine and thymine as pyrimidines. The same nucleotides occur in RNA except that thymine is replaced by uracil.
Organogenesis	The appearance, during development, of the organs characterizing the adult.
Phenotype	The observable characters, structural and functional, of an organism.
Plasmid	An extra-chromosomal hereditary determinant, consisting of DNA, found in bacteria.
Prokaryote	A cell lacking a membrane around its chromosomes, which are continuous, i.e., circular. (Also spelled procaryote.)
Protein	A large molecule made up of amino acids whose folding (conformation) determines its role as a structural and/or functional entity in a cell.
Protist	An **eukaryote** unicellular organism.
Somatic cell	Those cells of a multicellular organism that are not gametes, i.e., all cells other than those that form sperm or eggs.

β-Thalassemia   A hereditary disease resulting from a mutant gene that produces an abnormal hemoglobin causing a form of anemia.

Transcription   The formation of messenger RNA from DNA.

Translation   The formation of an amino acid chain (which will fold into a protein) as determined by messenger RNA.

A useful source of definitions relating to genetics is *A Dictionary of Genetics* by Robert C. King, Oxford University Press, New York, 2nd edition, 1972.

Index

E

D

F

G

H

I

R

Recombinant Advisory Committee,
   NIH 107, 111–112
Recombinant DNA
   companies 91–95
Recombinant DNA industry
   development 88–91
Recombinant DNA
   methods 15–18
Recombination 60
Recombination of DNA 60–61
Regulation 23, 51–53, 81–82,
              105–113, 117–18
Reproductive intervention 34–38
Research and
   economics 77–80, 90
Research-to-market time 92
Resins 30
Restriction of DNA 47–48
Restriction enzyme 16
RNA (ribonucleic acid) 11–14
RNA, comparison with DNA 11

S

Salt accumulation in soil 68–70
Salt-tolerant plants 70
Science and business 73–84
Science and society 115–119
Scientific community and
   Congress 106–107
Scientific and technical
   education 83–84
Secondary metabolic
   products 66–67
Seed companies 98–99
Selective breeding 24–25, 37
Self-regulation . . . . : . . 111–113
Serum albumin, human 95
Sickle cell anemia 14, 42–43
Society and science 115–119
Soil damage 68–70
Somatostation 79, 89, 93
Splicing of DNA molecules 16
Structure of DNA 6–10
Supreme Court rulings 21

Symbiosis, science and
   business 73–84
Synthesis gas 97
Synthetic organic chemicals,
   feedstock 30

T

Technical and scientific
   education 83–84
β-Thalassemia 38, 42
   See also Anemia
Thymine 8, 11
Thymosin alpha–1 93
Toxicity studies 63–64
Toxins, bacterial 93
Transcription 6, 11
Transduction 60
Transformation 60
Translation 6, 12
Translation, protein 13*f*
Tryptophan synthetase 61

U

United Nations task force 48
Universities and
   biotechnology 86–87
University research 55–56
Uracil 11
Urokinase 95

V

Vaccine 29
   hepatitis B 99
   hoof and mouth disease 95
Vaccine production 65
Viral diseases 79

W

Wood feedstock 97–98

129

Editing by L. Luan Corrigan
Production by Paula Bérard and Robin Giroux
Book design by Kathleen Schaner and Robin Giroux
Text illustrations by Martha Sewall

Typeset by Circle Graphics, Washington, DC, and Service
Composition Co., Baltimore, MD
Printed and bound by the Maple Press Co., York, PA